# Lecture Notes in Physics

**Editorial Board**

T0235698

# The Lecture Notes in Physics

The series Lecture Notes in Physics (LNP), founded in 1969, reports new developments in physics research and teaching – quickly and informally, but with a high quality and the explicit aim to summarize and communicate current knowledge in an accessible way. Books published in this series are conceived as bridging material between advanced graduate textbooks and the forefront of research to serve the following purposes:

• to be a compact and modern up-to-date source of reference on a well-defined topic;

• to serve as an accessible introduction to the field to postgraduate students and non-specialist researchers from related areas;

• to be a source of advanced teaching material for specialized seminars, courses and schools.

Both monographs and multi-author volumes will be considered for publication. Edited volumes should, however, consist of a very limited number of contributions only. Proceedings will not be considered for LNP.

Volumes published in LNP are disseminated both in print and in electronic formats, the electronic archive is available at springerlink.com. The series content is indexed, abstracted and referenced by many abstracting and information services, bibliographic networks, subscription agencies, library networks, and consortia.

Proposals should be sent to a member of the Editorial Board, or a Springer editor working in the Physics Editorial Department.

I.B. Khriplovich

# Theoretical Kaleidoscope

**Author**

I.B. Khriplovich
Budker Institute of Nuclear Physics
Novosibirsk 630090
Russia
E-mail: khriplovich@inp.nsk.su

I.B. Khriplovich, *Theoretical Kaleidoscope,* Lect. Notes Phys.
747 (Springer, New York 2008), DOI 10.007/978-0-387-75252-5

ISBN: 978-0-387-75251-8          e-ISBN: 978-0-387-75252-5

Library of Congress Control Number: 2007940954

Printed on acid-free paper.

9 8 7 6 5 4 3 2 1

springer.com

# Preface

This book is not a systematic course of theoretical physics, it contains only selected problems and topics of this remarkable science. One cannot but recall here the words of *I. Newton* (in his "Universal Arithmetic"): "In learning the Science, Examples are of more use than Precepts."

The book is based on lectures on additional chapters of theoretical physics given by me in various years at the Department of Physics, Novosibirsk University, Russia, and at Scuola Normale Superiore, Pisa, Italy. The main source of interest to the problems addressed in the lectures were my own investigations, as well as discussions with colleagues and students. The contents of the additional courses varied in time, together with my own interests, and sometimes in accordance with the wishes of the students. Some problems considered in the book were included also into general courses.

I strived to rely both in the lectures and in the book on qualitative, intuitive arguments, as far as they can be sufficient at all for theoretical physics. An exception in this sense is in the book the chapter on the semiclassical approximation in complex plane; it appears here by the following reason: in my opinion, a comprehensible presentation of this efficient technique is practically absent in literature. Unfortunately, in other chapters also, far from always I could avoid calculations.

I hope that the book will be accessible and useful for a student who has digested common courses of analytical mechanics, electrodynamics, and quantum mechanics; only some sections of the last chapter require an acquaintance with the fundamentals of quantum electrodynamics. Hopefully, even a mature physicist will find something interesting in the book. And not only a theorist. May be, the emphasis on a qualitative analysis of problems will make at least a major part of the book accessible for an experimentalist as well.

There is no list of references in the book. However, in the cases where, in my opinion, a reader may run into difficulties, references are given to the remarkable course of theoretical physics by L. D. Landau and E. M. Lifshitz; hopefully, these references will be useful and sufficient. In the book I indicate also the authors of concrete results.

Some results presented in the book were obtained in collaboration with A. D. Dolgov, D. V. Matvienko, E. V. Pitjeva, A. A. Pomeransky, and G. Yu. Ruban.

Useful remarks on the subjects discussed in the book were made by A. I. Chernykh, S. M. Kopeikin, A. A. Pomeransky, S. A. Rybak, V. G. Serbo, V. V. Sokolov, A. I. Vainshtein, and M. I. Vysotsky.

Of exceptional importance to me was the lively interest of numerous undergraduates and graduates.

I owe my deep and sincere gratitude to all of them.

Novosibirsk,                                                              *Iosif Khriplovich*
July 2007

# Contents

# Classical Mechanics. Unexpected Questions

## 1.1 What Is the Additional Integral of Motion for Isotropic Oscillator?

It is well-known that in the attractive Coulomb potential

$$U(r) = -\frac{\alpha}{r}, \quad \alpha > 0, \tag{1.1}$$

orbits of particles with negative energy are closed. Particle orbits are closed also in the isotropic oscillator potential

$$U(r) = \frac{1}{2}m\omega^2 r^2. \tag{1.2}$$

Let us note that, as has been proven long ago (*J. Bertrand, 1873*), there are no other central potentials, besides the Coulomb and oscillator ones, with motion along closed orbits [1]. To be more precise, we mean closed orbits which exist independently of the concrete values of two other, ordinary, integrals of motion, angular momentum $L$ and energy $E$. This remark is quite essential since, for instance, for a wide class of attractive potentials $U(r)$ closed circular orbits certainly exist. Their radius $r_0$ corresponds to the minimum of the effective potential

$$U_{\text{eff}}(r) = U(r) + \frac{L^2}{2mr^2}$$

of the radial motion, and is obviously a function of $L$. In its turn, the energy on such an orbit,

$$E = U_{\text{eff}}(r_0) = U(r_0) + \frac{L^2}{2mr_0^2},$$

depends on the angular momentum $L$, both explicitly and via $r_0(L)$.

---

[1] Unfortunately, the known proofs of this simple fact are rather tedious.

It is well-known that the fact that orbits for particles of negative energy in the attractive Coulomb potential are closed, is directly related to the existence in the problem, in line with the angular momentum integral $\mathbf{L}$, of one more conserved vector

$$\mathbf{A} = [\mathbf{v} \times \mathbf{L}] - \alpha \frac{\mathbf{r}}{r}, \qquad (1.3)$$

directed along the large semiaxis of the ellipse from the Coulomb center to the point on trajectory, which is of minimum distance from the center (*P. S. Laplace*, 1829; *K. Runge*, 1919; *W. Lenz*, 1924). One can easily check directly that vector (1.3) is an integral of motion, by differentiating it over time.

It is natural to assume that an oscillator potential, where trajectories are also closed, should also possess its own additional integral of motion. To find out what this integral looks like, we recall first of all an essential difference between closed elliptic orbits in the Coulomb and oscillator potentials.

In the first case, the Coulomb center, chosen as an origin, is in the focus of the ellipse, so that the particle trajectory is characterized indeed by a vector. This vector is directed from the center, for instance, to the point on the orbit of the shortest distance to the center.

In the second case, the oscillator one, the center of potential, i.e., the point where its value is minimum and which is chosen for the origin, is not in the focus of an elliptic trajectory, but in its center. It can be easily seen that here the trajectory has no singled out vector at all. In other words, if a trajectory in the Coulomb potential has one symmetry axis passing through the origin, in the oscillator problem there are two symmetry axes. Therefore, a particle trajectory in the oscillator potential may be characterized by a symmetric second-rank tensor.

One can easily check that tensor

$$Q_{mn} = \frac{m}{2} v_m v_n + \frac{m\omega^2}{2} r_m r_n, \qquad (1.4)$$

lying in the $xy$ plane ($z$ axis we direct along the orbital momentum $\mathbf{L}$), is an integral of motion indeed. After going over to main axes (of course, in the present case they coincide with the symmetry axes of an elliptic orbit in the oscillator potential), tensor (1.4) reduces to

$$Q_{mn} = \frac{m}{2} \, \mathrm{diag} \, (v_x^2 + \omega^2 x^2, \; v_y^2 + \omega^2 y^2) = \mathrm{diag} \, (E_x, \; E_y). \qquad (1.5)$$

Thus, conservation of tensor $Q_{mn}$ results from the fact that in an isotropic linear oscillator the energy is conserved for the motion along each of its axes separately.

In other words, if one identifies the coordinate axes with the symmetry axes of an elliptic orbit, the additional integral of motion in an oscillator potential reduces in fact to a single scalar function

$$E_- = E_x - E_y. \qquad (1.6)$$

In the Coulomb field, the situation is similar in this respect. Here also, if one directs one of the coordinate axes from the center to perihelion, the additional integral of motion reduces in fact to a single scalar function, the length of vector (1.3) (it can be easily seen that this length is equal to $\alpha\varepsilon$, where $\varepsilon$ is the orbit eccentricity).

We note that in both cases, Coulomb and oscillator ones, the existence of an additional integral of motion is related directly to the presence in each of them, in line with the spherical coordinates, of one more system where the variables can be separated. These are the parabolic coordinates in the Coulomb problem, and the common Cartesian coordinates in the oscillator one.

Let us recall now that in the Coulomb field vector (1.3) is conserved also for an infinite motion, i.e., in the case of positive energy, and even in the repulsive Coulomb field (for $\alpha < 0$). It can be easily checked, for instance, by direct calculation of $d\mathbf{A}/dt$, using equations of motion.

On the other hand, the existence of an additional conserved vector for an *infinite* motion in a central field is rather obvious and far from being special for the Coulomb problem (*I. B. Khriplovich, A. A. Pomeransky, G. Yu. Ruban,* 2006). Indeed, for a finite motion, the absence of precession from a turn to turn for the vector directed from the center to the point of trajectory which is closest to the center, is an exceptional property of the Coulomb field. But for an infinite motion, i.e., for a single passing by the center, such a vector is conserved trivially. However, just due to its triviality, this additional conserved vector gives here nothing new.

## 1.2 May Energy of a System Be Conserved If Its Hamiltonian Is Explicitly Time-Dependent?

To the surprise of some readers, the reply to the question asked in the title is positive [2].

Let us consider at first the well-known problem of a charged particle in a constant homogeneous magnetic field. The Hamiltonian of this particle is

$$H = \frac{1}{2m} \left( \mathbf{p} - \frac{e}{c}\mathbf{A} \right)^2. \tag{1.7}$$

It is well-known also that various gauges are possible for the vector potential $\mathbf{A}$. For magnetic field $\mathbf{B}$ directed along the $z$ axis, one can choose for instance

$$\mathbf{A} = B(0, x, 0). \tag{1.8}$$

In this gauge, the Hamiltonian is independent of $y$ and therefore the corresponding component $p_y$ of the canonical momentum is an integral of motion. One can choose, however, another gauge:

---

[2]In the presentation here, we follow the note by *I. B. Khriplovich, A. I. Milstein* (1999).

$$\mathbf{A} = B(-y, 0, 0). \tag{1.9}$$

And then another component of the canonical momentum, $p_x$, will be conserved.

Thus, a component of $\mathbf{p}$, orthogonal to the magnetic field, has turned out an integral of motion, and moreover, the conserved component of momentum can be chosen at will. What does it mean?

A sufficiently obvious answer is that the canonical momentum $\mathbf{p}$ is not a gauge-invariant quantity, and hence has no direct physical meaning. As to the usual visual picture of the transverse motion in magnetic field, it is not the canonical momentum $\mathbf{p}$ that precesses and changes constantly its direction, but the velocity

$$\mathbf{v} = \frac{1}{m} \left( \mathbf{p} - \frac{e}{c} \mathbf{A} \right).$$

As distinct from the canonical momentum $\mathbf{p}$, the velocity $\mathbf{v}$ is a gauge-invariant and uniquely defined quantity.

It is quite natural that not only the space component $\mathbf{p}$ of canonical momentum is gauge-dependent, but its time component, the Hamiltonian $H$, is gauge-dependent as well. It is the kinetic energy $H - eA_0$ which is gauge-invariant.

As a rather unexpected manifestation of this fact, let us consider the example of a well-known physical system where the energy is conserved, but the Hamiltonian may be time-dependent. We mean a charged particle in an electric field $\mathbf{E}$, for instance, in a Coulomb one. Let us choose here the gauge $A_0 = 0$. Obviously, in this gauge the vector potential becomes equal to $\mathbf{A} = -ct\,\mathbf{E}$, so that now Hamiltonian (1.7) is explicitly time-dependent.

Nevertheless, the energy of a particle in a time-independent electromagnetic field is certainly conserved. Indeed, here the equations of motion are

$$\dot{\mathbf{r}} = \{H, \mathbf{r}\} = \frac{1}{m} \left( \mathbf{p} + e\,t\mathbf{E} \right), \tag{1.10}$$

$$m\ddot{\mathbf{r}} = \frac{d}{dt} \left( \mathbf{p} + e\,t\,\mathbf{E} \right) = \frac{\partial}{\partial t} \left( \mathbf{p} + e\,t\,\mathbf{E} \right) + \{H,\ \mathbf{p} + e\,t\,\mathbf{E}\} = e\mathbf{E} \tag{1.11}$$

(we use in these classical equations the Poisson brackets $\{..., ...\}$). Since the strength of a time-independent electric field can always be written as the gradient of some scalar function:

$$\mathbf{E} = -\nabla\phi,$$

equation (1.11) has first integral

$$\frac{1}{2} m\dot{\mathbf{r}}^2 + e\phi = \text{const},$$

which is nothing but the energy integral.

On the other hand, due to equation (1.10), Hamiltonian in the gauge $A_0 = 0$ coincides in fact with the *kinetic* energy:

$$H = \frac{1}{2m} (\mathbf{p} + e t \mathbf{E})^2 = \frac{1}{2} m \dot{\mathbf{r}}^2. \qquad (1.12)$$

It should be the case since the kinetic energy $H - eA_0$, being a gauge-invariant quantity, should coincide with the Hamiltonian in the gauge $A_0 = 0$.

## 1.3 Dark Matter in Solar System, Gauss Theorem, and Secular Precession of Planet Perihelion

Astronomical observations indicate that in the universe, along with the common matter, there is a so-called dark matter, which interacts with the common one only gravitationally. Moreover, the amount of the dark matter is larger than that of the common one.

The average density of the dark matter in the universe constitutes about

$$\rho_{\text{uni}} \sim 10^{-29} \text{ g/cm}^3,$$

and its density in our galaxy is much higher:

$$\rho_{\text{gal}} \sim 10^{-24} \text{ g/cm}^3.$$

As to the dark matter density in our solar system, only upper limits on it are known. Even the best of them, discussed below, correspond to a density much higher than the galactic one. It is natural to ascribe the possibility of existence of such relatively high density to the gravitational field of the Sun. This justifies, at least partially, the assumption made below, according to which the dark matter in our solar system is a dust with density $\rho(r)$, spherically symmetric with respect to the Sun.

Let us find the gravitational potential $\Phi(r)$ of such a dust. It is instructive to use to this end the Gauss theorem for the gravitational field strength, i.e., for the acceleration $g(r)$ created by this field:

$$g(r) = -\frac{4\pi k}{r^2} \int_0^r dr_1 \, r_1^2 \, \rho(r_1); \qquad (1.13)$$

here $k$ is the Newton gravitational constant. Let us attract attention to the sign $-$ in this expression, which corresponds to the fact that the acceleration is directed to the center, but not in the opposite direction, for any positively defined density $\rho$. The gravitational potential is [3]

---

[3] One should be warned against a possible naïve (and erroneous!) presentation of the Gauss theorem in the form $\Phi(r) = -k\mu(r)/r$, where $\Phi(r)$ is the gravitational potential, and $\mu(r) = 4\pi \int_0^r dr_1 \, r_1^2 \, \rho(r_1)$ is the total mass of the matter inside the sphere of radius $r$. Obviously, if $\mu(r)$ grows with radius faster than $r$, such potential would result in antigravity, i.e., in the gravitational repulsion, but not attraction.

$$\Phi(r) = -\int^r dr_2\, g(r_2) = 4\pi k \int^r dr_2\, \frac{1}{r_2^2} \int_0^{r_2} dr_1\, r_1^2\, \rho(r_1) \,. \qquad (1.14)$$

As usual, a potential is defined up to a constant, so that the value of the lower limit in the integral over $r_2$ is inessential. Of course, by changing the order of integration, this formula reduces to the common expression for the potential created by a spherically symmetric density:

$$\Phi(r) = -4\pi k \int \frac{dr'\rho(r')}{|\mathbf{r} - \mathbf{r}'|} = -4\pi k \left[ \frac{1}{r} \int_0^r dr'\, r'^2 \rho(r') + \int_r^\infty dr'\, r' \rho(r') \right].$$

The corresponding correction to the potential energy of a planet with mass $m$ is $\delta U(r) = m\, \Phi(r)$. This correction shifts the perihelion of a planet orbit by the angle [4]

$$\delta\phi = \frac{d}{dL}\, \frac{2m}{L} \int_0^\pi d\phi\, r^2\, \delta U(r) \qquad (1.15)$$

per period. It is convenient to go over in this expression from the orbital angular momentum $L$ to the so-called orbit parameter $p$; they are related as follows:

$$p = \frac{L^2}{k\, m^2 M}\,; \qquad (1.16)$$

here and below $M$ is the mass of the Sun. Besides, we express $r$ via $p$ and the orbit eccentricity $e$:

$$r = \frac{p}{1 + e \cos\phi}\,. \qquad (1.17)$$

Then, the relative perihelion shift per period is written as

$$\frac{\delta\phi}{2\pi} = \frac{1}{kmM} \left( -\frac{1}{p} + 2\frac{d}{dp} \right) \int_0^\pi \frac{d\phi}{\pi} R\left( \frac{p}{1 + e\cos\phi} \right), \qquad (1.18)$$

where $R(r) = r^2\, \delta U(r)$.

We note that for the planets of interest to us the eccentricities are small: for Mercury, Earth, and Mars they are 0.21, 0.02, and 0.09, respectively. We use this circumstance, and expand the integrand in $e$, up to second order included. Then, integration over $\phi$ results in

$$\frac{\delta\phi}{2\pi} = \frac{1}{kmM} \left( -\frac{1}{p} + 2\frac{d}{dp} \right) \left\{ R(p) + \frac{1}{2}e^2 \left[ p\, R'(p) + \frac{1}{2}p^2 R''(p) \right] \right\}. \quad (1.19)$$

Now we put $e^2 = 1 - 2p|E|/(kmM)$ equal to zero, with the account for relation

$$\frac{de^2}{dp} = -\frac{2|E|}{(kmM)} \longrightarrow -\frac{1}{r_0}\,,$$

where $r_0$ is the radius of a circular orbit. From now on we omit the subscript 0 at this radius, and arrive at expression

---

[4]See: L. D. Landau and E. M. Lifshitz, *Mechanics*. §15 (in particular, Problem 3).

$$\frac{\delta\phi}{2\pi} = \frac{1}{kmM}\left[-\frac{1}{r}R(r) + R'(r) - \frac{1}{2}r^2 R''(r)\right]. \tag{1.20}$$

Coming back in it from $R(r)$ to $r^2\,\delta U(r)$, we arrive finally at a simple result:

$$\frac{\delta\phi}{2\pi} = -\frac{2\pi}{M}\,\rho(r)\,r^3. \tag{1.21}$$

One can arrive at this result otherwise, by calculating the following corrections caused by the perturbation $\delta U(r)$: $\delta\omega_\phi$ to the rotation frequency and $\delta\omega_r$ to the frequency of small radial oscillations with respect to the circular orbit. The difference of these corrections, multiplied by the unperturbed rotation period, is equal to the perihelion shift per period. The corresponding calculation is no less tedious than that given above. The advantage of the derivation presented in detail here is that, if necessary, it allows one to include in elementary way corrections due to the finite eccentricity. However, it is clear from the above solution that these corrections start at second order in $e$. Therefore, even for Mercury, with the largest eccentricity, 0.21, these corrections are inessential for the problem under discussion.

Coming back to formula (1.21), we wish to emphasize that, according to it, the perihelion shift is governed directly by a local property of dark matter, i.e., by its density $\rho(r)$ on the trajectory of a planet. Therefore, the analysis of observational data for the secular perihelion precession of various planets results in direct, model-independent upper limits on the local density of dark matter at various distances from the Sun, corresponding to the orbit radii.

In Table 1.1 we present these limits (*I. B. Khriplovich, E. V. Pitjeva,*

*Table 1.1*

|  | Mercury | Earth | Mars |
|---|---|---|---|
| excessive perihelion shift, $10^{-4}$ s/century | $-0.67\pm0.93$ | $-0.15\pm0.31$ | $0.14\pm0.73$ |
| orbit radius $r$, a.u. | 0.39 | 1.00 | 1.52 |
| dark matter density, $10^{-19}$ g/cm$^3$ | $110\pm150$ <br> $<260$ | $1.4\pm3.0$ <br> $<4.4$ | $-0.4\pm2.0$ <br> $<1.6$ |

2006), following from the analysis of the perihelion precession of Mercury, Earth, and Mars. In the table, the excessive perihelion shift means a possible correction to the secular perihelion precession, i.e., deviation of the most accurate calculations (*E. V. Pitjeva*, 2005) from observational data; just for the planets considered here these deviations are minimal. Then, the average orbit radii for these planets are indicated in astronomic units; 1 a.u. = 150×10^6 km. The upper limits on the dark matter density in the last line of the table are derived in an obvious naïve way from the numbers in the previous line.

## 1.4 What Is Classical Analogue of Schrödinger Variational Principle?

### 1.4.1 General Discussion

Let us recall the formulation of the Schrödinger variational principle in quantum mechanics. For the sake of simplicity, we confine ourselves here to the one-dimensional bound-state problem. The wave function $\psi(x)$ of a system with Hamiltonian $H$ is found by the variational method as a minimum of the functional

$$\int dx\, \psi(x) H \psi(x)\,, \tag{1.22}$$

at the subsidiary normalization condition

$$\int dx\, \psi^2(x) = 1 \tag{1.23}$$

(it is well-known that one can choose here real wave functions). Obviously, the success of the procedure depends essentially on judicious choice of the trial function $\psi(x)$. In particular, the ground state wave function $\psi_0(x)$ should not have nodes. Indeed, just this condition guarantees the minimum value of the mean kinetic energy

$$-\frac{\hbar^2}{2m} \int dx\, \psi_0(x) \Delta \psi_0(x) = \frac{\hbar^2}{2m} \int dx\, \left(\frac{d\psi_0(x)}{dx}\right)^2.$$

After $\psi_0(x)$ has been found, the wave function $\psi_1(x)$ of the first excited state is obtained also as a minimum value of functional (1.22) with the normalization condition (1.23). However, $\psi_1(x)$ should be orthogonal to $\psi_0(x)$, i.e., it should satisfy one more subsidiary condition

$$\int dx \psi_0(x) \psi_1(x) = 0\,. \tag{1.24}$$

In its turn, wave function $\psi_1(x)$ should also have the minimum possible number of nodes. Since $\psi_1(x)$ is orthogonal to the ground state function $\psi_0(x)$ of definite sign, then (again to minimize the kinetic energy) the wave function $\psi_1(x)$ of the first excited state should have one node.

Analogously, the wave function $\psi_2(x)$ of the second excited state should be orthogonal both to $\psi_0(x)$ and $\psi_1(x)$. Correspondingly, it has two nodes.

Sufficiently evident general rule is as follows: the wave function $\psi_n(x)$ of the $n$th excited state has $n$ nodes. In other words, the Schrödinger variational principle in quantum mechanics can be formulated for the $n$th state as the requirement of minimum average energy at the fixed number $n$ of the nodes.

Let us recall now the semiclassical Bohr–Sommerfeld quantization rule

$$\int_a^b p\,dq = \pi\hbar\left(n + \frac{1}{2}\right). \tag{1.25}$$

Then it becomes clear that in the classical limit of large $n$, the Schrödinger variational principle is formulated as the requirement of minimum average energy $\bar{E}$ at the fixed truncated action

$$W = \int_a^b p\,dq\,,$$

i.e.,

$$\delta\bar{E}\,|_W = 0\,. \tag{1.26}$$

On the other hand, in classical mechanics there is the Maupertuis principle, which is formulated as follows. Let us write the action for the trajectories where the energy $E$ is conserved in the form

$$S = \int_a^b p\,dq - Et = W - Et\,. \tag{1.27}$$

The Maupertuis principle consists in the requirement of minimum truncated action $W$ at fixed energy $E$,

$$\delta W\,|_E = 0\,. \tag{1.28}$$

Of course, at the considered trajectories the mean energy $\bar{E}$ coincides with $E$, so that the classical analogue (1.26) of the Schrödinger variational principle is in fact reciprocal one with respect to the Maupertuis principle (1.28) (*C. Grey, G. Karl, V.A. Novikov*, 1996). Certainly, variational principle (1.26) applies not only to a finite motion, but to infinite motion as well.

### 1.4.2 Example. Anharmonic Oscillator

As an example of the efficient application of variational principle (1.26), we consider the classical problem of the anharmonic oscillator with Hamiltonian

$$H = \frac{p^2}{2m} + \frac{1}{2}m\omega_0^2 x^2 + \frac{1}{4}\beta m x^4. \tag{1.29}$$

As well as in quantum mechanics, the success of the direct variational approach to a classical problem depends essentially on the judicious choice of the trial function.

In spite of the presence of nonlinear term in Hamiltonian (1.29), the particle motion remains periodic. Therefore, we should choose the trial solution in a periodic form as well. Let us write it as

$$x(t) = a \sin \omega t. \tag{1.30}$$

Then the truncated action and mean energy are, correspondingly,

$$W = \int_0^T p\,dx = \int_0^T dt\, m\dot{x}^2 = ma^2\omega\, \frac{1}{2} \int_0^{2\pi} d\phi = \pi ma^2\omega = \text{const}, \tag{1.31}$$

and

$$\bar{E} = \frac{1}{T} \int_0^T dt \left[ \frac{1}{2} \left( \frac{p^2}{m} + m\omega_0^2 x^2 \right) + \frac{1}{4} m\beta x^4 \right]$$

$$= \frac{ma^2}{4} \left[ \omega^2 + \omega_0^2 + \frac{3}{8}\beta a^2 \right] = \frac{W}{4\pi} \left[ \omega + \frac{\omega_0^2}{\omega} + \frac{3}{8} \frac{\beta W}{\pi m\omega^2} \right]; \tag{1.32}$$

here $T = 2\pi/\omega$.

It can be easily seen that, with a constant $W$, the minimum of $\bar{E}$ in $\omega$ is reached for

$$\omega^2 = \omega_0^2 + \frac{3}{4} \frac{\beta W}{\pi m\omega}. \tag{1.33}$$

With formulas (1.31) and (1.33) one can in principle express both the frequency $\omega$ and amplitude $a$ of a nonlinear oscillation via the parameters of Hamiltonian (1.29) and the fixed value of truncated action $W$. Thus, these formulas give in an implicit form the complete variational solution of the formulated problem.

Let us rewrite relation (1.33) somewhat otherwise:

$$\omega^2 = \omega_0^2 + \frac{3}{4}\beta a^2. \tag{1.34}$$

It becomes obvious now that the frequency of nonlinear oscillations depends on their amplitude, as distinct from the linear case. We note also that in the limit of weak nonlinearity, relation (1.34) reproduces the well-known perturbative result for the frequency shift of a nonlinear oscillator:

$$\omega = \omega_0 + \frac{3}{8} \frac{\beta a^2}{\omega_0}. \tag{1.35}$$

# 2

# Wave Phenomena and Classical
# Electrodynamics Without Calculations

## 2.1 Uncertainty Relation, Diffraction,
## and Narrow Waveguides

Let us consider the well-known problem of diffraction of an initially plane
wave with wave vector $k$ on a round hole of radius $a$ in a thin screen. If the
radius of the hole is sufficiently large, so that $ka \gg 1$, then the wave remains
essentially plane after going through the hole, with a small distortion due to
a diffraction on the edges of the hole. Now we start diminishing $a$. The wave
gets more and more distorted after going through the hole. Indeed, the allowed
transverse component of the wave vector in it increases in accordance with
the uncertainty relation $\Delta k_t a \gtrsim 1$. At last, with $a \sim 1/k$ (or $a \sim \lambda$, where $\lambda$ is
the wavelength) the outgoing wave becomes spherical, since in this case the
transverse component of the wave vector in it $k_t \gtrsim 1/a$ reaches its maximum
allowed value $k$.

However, let us decrease further the size of the hole. Obviously, on the
one hand, $k_t$ cannot exceed $k$, but on the other hand, one cannot violate the
uncertainty relation. So, what will happen?

The first answer is that with too small a hole the wave just will not pen-
etrate it. This is true by itself in the sense that for $a \ll \lambda$ the amplitude of
the outgoing wave will be exponentially small. In principle, however, one can
make, first, the intensity of the incoming wave arbitrarily large, and second,
the detector behind the screen arbitrarily sensitive. Hence, this answer does
not solve the problem.

The solution looks otherwise. The uncertainty relation dictates indeed that
for the wave with wave vector $k$ the size of its source should not be less than
$\lambda \sim 1/k$. When applied to our problem, it means that the source of the
outgoing wave is in fact not the hole of radius $a$ by itself, but a spot on the
external side of the screen with a characteristic size $\sim \lambda \gg a$. The quantitative
solution of the problem of wave diffraction on a small hole is too complicated
to be considered here. However, the pointed out qualitative answer is clear
and dictated just by the uncertainty relation.

And now we somewhat modify the problem. Instead of a thin screen with a round hole, let us consider a screen of a finite thickness with a round channel. We assume that the walls of the channel are absolutely reflecting, and its radius $a$ is small, i.e., again $ka \ll 1$. In other words, we consider the passage of a wave through a narrow waveguide. In such a channel (or wave guide), the wave certainly cannot spread in the transverse direction. Then, how can one reconcile condition $ka \ll 1$ with the uncertainty relation $k_t a \gtrsim 1$?

The answer is as follows. The components of the wave vector, longitudinal $k_l$ and transverse $k_t$, are related by the obvious relation

$$k_l^2 + k_t^2 = k^2.$$

Therefore, the fact that in our narrow wave guide $k_t^2 \gg k^2$ means that in it $k_l^2 < 0$. In other words, in such a wave guide the longitudinal component $k_l$ of wave vector is purely imaginary. And this corresponds of course to the well-known fact: a wave in a narrow wave guide is damped exponentially.

## 2.2 Pseudoscalar Invariant of Electromagnetic Field. Particle with Magnetic and Electric Dipole Moments

It is well-known that an electromagnetic field is characterized by two quadratic invariants

$$I_1 = F_{\mu\nu}F_{\mu\nu} = 4(\mathbf{B}^2 - \mathbf{E}^2), \quad I_2 = \frac{1}{2}\varepsilon_{\alpha\beta\mu\nu}F_{\alpha\beta}F_{\mu\nu} = 4\,\mathbf{E}\mathbf{B};$$

here $F_{\mu\nu} = \partial_\mu A_\nu - \partial_\nu A_\mu$ is the tensor of field strength, $A_\mu$ is its vector potential, Greek indices run the values 0, 1, 2, 3. The electric field strength $\mathbf{E}$ is a common polar vector, besides it does not change sign under time reversal. The magnetic field strength $\mathbf{B}$ is an axial vector, and changes sign under time reversal. Correspondingly, $I_1$ is a true scalar, and $I_2$ is a pseudoscalar changing sign under time reversal.

It can be easily seen that the pseudoscalar $I_2$ is in the general case a total divergence:

$$I_2 = \partial_\alpha(\varepsilon_{\alpha\beta\mu\nu}A_\beta F_{\mu\nu}).$$

In particular, for static fields $I_2$ reduces to the divergence of a three-dimensional vector:

$$I_2 = 4\,\mathbf{E}\mathbf{B} \longrightarrow -4\,\mathbf{B}\,\boldsymbol{\nabla}A_0 = -4\,\boldsymbol{\nabla}(A_0\mathbf{B}). \qquad (2.1)$$

Therefore, if static fields decrease sufficiently fast at infinity, the integral of $I_2$ over the whole space vanishes.

Let us consider now a particle which possesses simultaneously both magnetic $\boldsymbol{\mu}$ and electric $\mathbf{d}$ dipole moments, parallel (or antiparallel) to each other.

For instance, it can be the neutron, a particle with spin $s = 1/2$, possessing a magnetic moment. Neutron is a nongenerate quantum-mechanical system,

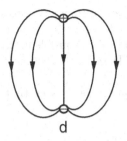 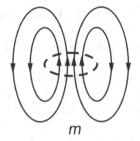

*Fig. 2.1*

characterized by a single vector, its spin **s**. Therefore, the neutron magnetic moment $\mu$ should be directed parallel or antiparallel to its spin (in the neutron they are in fact antiparallel). We note that both $\mu$ and **s** are axial vectors. Besides, they both change sign under the time reversal. There are good reasons to believe that the neutron may possess also an electric dipole moment (EDM) **d**. By the exactly same reason, the absence of degeneracy, **d** should be directed parallel or antiparallel to the spin **s**. However, as distinct from the spin and magnetic moment, the EDM **d** is a polar vector, and besides, it does not change sign under time reversal. Therefore, the statement that **d** and **s** are parallel (or antiparallel), changes to the opposite one both under the reflection of space coordinates and under the time reversal, i.e., this statement is not invariant with respect to both these operations. Thus, the existence of electric dipole moments of the neutron (as well as other elementary particles) is forbidden by the laws of conservation of space and time parity. However, just by this reason, the searches for the dipole moments are of great interest for elementary particle physics. Still, the problems of violation of space and time parity are not directly related to the subject of the present section. We come back therefore to our problem.

Let us, for definiteness sake, assume that $\mu$ and **d** are parallel. Then, it seems that magnetic and electric lines of strength are also parallel everywhere. In this case the invariant $I_2$ is positively definite, so that the integral of it over the whole space cannot turn to zero. How is this paradox resolved [1] ?

Let us consider the picture of lines of strength in more details, taking into account the finite size of the dipoles themselves. From inspecting Fig. 2.1 it becomes clear that if **d** and $\mu$ are parallel, and correspondingly, the electric and magnetic lines of strength are parallel in the outer region, then the same lines of strength are antiparallel inside the dipoles. The contributions of the outer

---

[1] Curiously, its resolution, at least in the classical version of the problem, does not create usually great difficulties for qualified experimentalists. On the other hand, this question sometimes turns out very difficult for quite well-known theorists.

and inner regions to the discussed integral cancel, in complete correspondence with the made assertion.

Let us address now the quantum version of the problem, where both dipoles, $\boldsymbol{\mu}$ and $\mathbf{d}$, are considered as point-like. Here some refinement of widely used formulas will be necessary.

We start with expression $\nabla_i \nabla_j (1/r)$. It can be easily seen that the naïve (and commonly used) relation

$$\nabla_i \nabla_j \frac{1}{r} = \frac{3r_i r_j - \delta_{ij} r^2}{r^5}$$

is generally speaking wrong. Indeed, it is sufficient to calculate its trace and recall that $\Delta(1/r) = -4\pi\delta(\mathbf{r})$. Obviously, the correct equation is as follows:

$$\nabla_i \nabla_j \frac{1}{r} = \frac{3r_i r_j - \delta_{ij} r^2}{r^5} - \frac{4\pi}{3}\delta_{ij}\,\delta(\mathbf{r})\,. \tag{2.2}$$

Let us find the magnetic field created by a point-like magnetic dipole moment $\boldsymbol{\mu}$. It is well-known that the corresponding vector potential equals

$$\mathbf{A} = \boldsymbol{\mu} \times \frac{\mathbf{r}}{r^3} = \nabla\frac{1}{r} \times \boldsymbol{\mu}\,. \tag{2.3}$$

And the magnetic field of this dipole is [2]

$$\mathbf{B} = \nabla \times \mathbf{A} = \frac{3\mathbf{n}(\mathbf{n}\boldsymbol{\mu}) - \boldsymbol{\mu}}{r^3} + \frac{8\pi}{3}\boldsymbol{\mu}\,\delta(\mathbf{r})\,, \quad \mathbf{n} = \frac{\mathbf{r}}{r}\,. \tag{2.4}$$

Now we consider the electric dipole moment $\mathbf{d}$. Its electric field is

$$\mathbf{E} = (\mathbf{d}\nabla)\nabla\frac{1}{r} = \frac{3\mathbf{n}(\mathbf{n}\mathbf{d}) - \mathbf{d}}{r^3} - \frac{4\pi}{3}\mathbf{d}\,\delta(\mathbf{r})\,. \tag{2.5}$$

Let us compare now expressions (2.4) and (2.5). For parallel $\mathbf{d}$ and $\boldsymbol{\mu}$, the common, long-range (i.e., without $\delta$-functions) contributions in these field strengths are also parallel. As to the local, $\delta$-function contributions, they are antiparallel for parallel $\mathbf{d}$ and $\boldsymbol{\mu}$. The correspondence is obvious with the structure of the lines of field strength of classical dipoles (see Fig. 3.1).

## 2.3 Synchrotron Radiation of Ultrarelativistic Particles Without Special Functions

The synchrotron radiation, i.e., the radiation of a charged particle in an external magnetic field, is considered in various textbooks [3]. However, the consideration therein is based usually on the rather tedious exact solution of the

---

[2] Let us note that just the $\delta$-function term in this expression is responsible for the hyperfine splitting of $s$-levels in atoms and ions caused by the nuclear magnetic moment.

[3] See, for instance: L.D. Landau and E.M. Lifshitz, *The Classical Theory of Fields*. §74.

problem. Meanwhile, the qualitative investigation, presented below, allows one to obtain in an intuitive way all principal characteristics of the synchrotron radiation of ultrarelativistic particles: its total intensity, angular and spectral distributions.

Let us start with the total radiation intensity. In the locally inertial frame (LIF), comoving with an electron, it is

$$I' \sim e^2 (a')^2 \sim \frac{e^4}{m^2} (E')^2. \tag{2.6}$$

Here $e$ and $m$ are the electron charge and mass, $a$ is its acceleration, $E$ is the electric field strength; $I$, $a$, and $E$ are supplied with primes to point out that they refer to the LIF. $E'$ is obtained from the magnetic field $B$ in the laboratory frame (LF) by the Lorentz transformation

$$E' \sim B\gamma, \quad \gamma = \frac{1}{\sqrt{1 - v^2}}. \tag{2.7}$$

We recall now that $I$ is an invariant. Indeed, the radiation intensity is expressed through the probability of the photon emission $W$ and its energy $\hbar\omega$ as follows: $I = W\hbar\omega$. Then, the probability $W$ in the LF is related to the probability in the LIF $W'$ by the relation $W = W'/\gamma$ (just recall that the lifetime of an unstable particle in LF is $\gamma$ times larger than that in LIF). On the other hand, it is well-known that $\omega = \omega'\gamma$. Finally, $I' = I$.

Now, substituting into (2.6) expression (2.7) for the electric field $E'$ in the LIF, we obtain the well-known result for the total intensity of radiation

$$I \sim \frac{e^4}{m^2} B^2 \gamma^2. \tag{2.8}$$

If instead of $B$ one fixes the radius of the electron trajectory $r_0$, related to $B$ via $eB \sim m\gamma/r_0$, the expression for the total intensity becomes

$$I \sim \frac{e^2 \gamma^4}{r_0^2}. \tag{2.9}$$

Let us go over now to the angular distribution of the radiation. In the LIF it has a common dipole form, it is just trigonometry. In other words, in the LIF $\theta' = k_t'/k_l' \sim 1$. Here $k_{t(l)}'$ is the transverse (longitudinal) component of the wave vector of the photon. In the LF these components are: $k_t = k_t'$, $k_l = k_l'\gamma$. Therefore, in the LF an ultrarelativistic electron radiates into a cone with a typical angle

$$\theta_c \sim k_t/k_l \sim \gamma^{-1}. \tag{2.10}$$

An observer receives the signal only staying inside this cone which moves together with the electron. An elementary consideration demonstrates that the electron beams at the observer only from the piece of the trajectory arc that has the same angular size as the cone itself. In the present case it means that the angular size of this piece of the arc is $\theta_c \sim \gamma^{-1}$. In other words, the formation length for radiation, which in our ultrarelativistic case ($v \approx c = 1$) coincides with the formation time for it, is

$$\Delta t \sim r_0 \theta_c \sim r_0 \gamma^{-1}.$$

Then the duration of signal receiving, with the account for the longitudinal Doppler effect, is

$$\delta t = (1 - \mathbf{n}\mathbf{v})\Delta t \approx \frac{1}{2}(\theta^2 + \gamma^{-2})\Delta t, \tag{2.11}$$

where $\mathbf{n} = \mathbf{k}/k$. For $\theta \sim \theta_c \sim \gamma^{-1}$ we obtain $\delta t_c \sim r_0 \gamma^{-3}$. It means that the characteristic frequency of the received radiation is $\gamma^3$ times larger than the rotation frequency $\omega_0$:

$$\omega_c \sim \delta t_c^{-1} \sim \gamma^3 r_0^{-1} \sim \gamma^3 \omega_0. \tag{2.12}$$

We turn now to the spectral distribution of the synchrotron radiation. Its intensity decreases rapidly for $\omega \gg \omega_c$. Let us assume that for $\omega \lesssim \omega_c$ it changes according to a power law: $I(\omega) \sim \omega^\nu$. Then, by comparing the total intensity given by the integral

$$\int^{\omega_c} d\omega I(\omega) \sim \omega_c^{\nu+1} \sim \gamma^{3(\nu+1)}$$

with formula (2.9), we obtain $\nu = 1/3$. In other words,

$$I(\omega) \sim \omega^{1/3} \quad \text{for} \quad \omega \lesssim \omega_c, \tag{2.13}$$

or for the discrete spectrum

$$I_n \sim n^{1/3} \quad \text{for} \quad n \lesssim \gamma^3. \tag{2.14}$$

And at last, let us find the angular distribution of radiation for the frequency range

$$\omega_0 \ll \omega \ll \omega_c, \qquad 1 \ll n \ll \gamma^3.$$

It is natural to expect that here the characteristic angles $\theta$ are larger than $\gamma^{-1}$. As previously, while the angle of the radiation cone is small, $\theta \ll 1$, the electron beams at the observer only from the piece of the trajectory arc which has the same angular size $\theta$. But then, instead of relation (2.11), we obtain

$$\delta t \sim \omega^{-1} \sim \theta^2 \Delta t \sim \theta^3 r_0 \sim \theta^3 \omega_0^{-1}.$$

Thus, in this frequency region

$$\theta \sim \left(\frac{\omega_0}{\omega}\right)^{1/3} \sim n^{-1/3}. \tag{2.15}$$

In the conclusion of this section, it should be emphasized that the obtained qualitative results are not special for the considered problem of finite motion of an ultrarelativistic particle in a magnetic field. They are applicable as well to a more general case, that of scattering in external electromagnetic fields if characteristic scattering angles exceed $\gamma^{-1}$.

## 2.4 How Does Front of Electromagnetic Wave Propagate in Medium?

We discuss in this section the propagation of a wave packet in a medium with frequency-dependent refraction index $n(\omega)$. Of course, the maximum of a wave packet propagates with the group velocity. However, we will be interested here in the velocity of wave front propagation. It is convenient to start the discussion of this question from another problem, so much the more that the problem is of independent interest [4].

### 2.4.1 Causality and Analyticity

Let us consider a system (for instance, a tuning fork) which transforms a received time-dependent signal $f(t)$ (a packet of sound waves), into a response $g(t)$ (sound of the tuning fork itself). Let also the response be related to the signal in a linear way:

$$g(t) = \int dt' L(t, t') f(t') . \tag{2.16}$$

If the tuning fork properties are independent of time, then the response function $L$ depends on the difference $t - t'$ only, so that

$$g(t) = \int dt' L(t - t') f(t') . \tag{2.17}$$

Let the signal be of the form $f(t) = \delta(t)$. Then at the exit we obtain

$$g(t) = L(t) .$$

Of course, the tuning fork cannot start vibrating before the moment $t = 0$, this is in fact required by causality. Therefore, it follows from the causality condition that

$$L(t) = 0 \quad \text{for} \quad t < 0 .$$

This evident condition results in quite nontrivial consequences for analytic properties of the Fourier transform $L(\omega)$ of the response function in the complex plane $\omega$. Indeed, the integral defining this Fourier transform,

$$L(\omega) = \int_{-\infty}^{\infty} dt\, e^{i\omega t}\, L(t) = \int_{0}^{\infty} dt\, e^{i\omega t}\, L(t) , \tag{2.18}$$

for any reasonable behavior of $L(t)$ at $t \to \infty$, converges everywhere in the upper half-plane $\omega$, i.e., it has no singularities therein. In other words, the function $L(\omega)$ is analytic in the upper half-plane.

---

[4] In the discussion of this problem we essentially follow *R. Hagedorn* (1966).

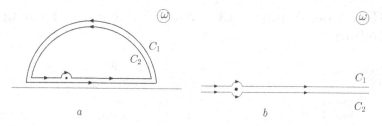

Fig. 2.2

From the analyticity in the upper half-plane of a response function, an important relation follows for its real and imaginary parts. Let a point $\omega$ be in the upper half-plane slightly above the real axis (see Fig. 2.2, $a$). We write the Cauchy formula for the function $L(\omega)$,

$$L(\omega) = \frac{1}{2\pi i} \int_{C_a} \frac{d\omega' L(\omega')}{\omega' - \omega} + \frac{1}{2\pi i} \int_{C_b} \frac{d\omega' L(\omega')}{\omega' - \omega} , \tag{2.19}$$

choosing the contours $C_1$ and $C_2$ as indicated in Fig. 2.2, $a$. Of course, the integral over the contour $C_2$ vanishes, it is included in formula (2.19) only for convenience. We neglect now the contributions from the integration over the semicircles of a large radius, which decrease exponentially with this radius, and put the lower horizontal line slightly below the real axis (see Fig. 2.2, $b$). As a result, we arrive at the so-called dispersion relation:

$$L(\omega) = \frac{1}{\pi i} P \int_{-\infty}^{\infty} \frac{d\omega' L(\omega')}{\omega' - \omega} , \tag{2.20}$$

where the integral is taken just along the real axis, and the symbol $P$ means, as usual, the principal value of integral. In particular, the real part of this relation looks as follows:

$$\mathrm{Re}L(\omega) = \frac{1}{\pi} P \int_{-\infty}^{\infty} \frac{d\omega' \, \mathrm{Im}L(\omega')}{\omega' - \omega} . \tag{2.21}$$

Let us demonstrate the application of dispersion relation (2.21) with the following example, formulated for the sake of definiteness in the language of quantum mechanics (though it refers, in essence, to a much wider class of problems). Let a wave be scattered off some potential. The scattering amplitude is nothing but the function of response by a potential to an incoming wave, and satisfies the same relation (2.21). In the nonrelativistic scattering problem, the particle energy $E$ plays the part of the wave frequency $\omega$. It is known that, according to the optical theorem, the imaginary part of the forward scattering amplitude $A(E)$ is related to the total scattering cross-section as follows:

$$\mathrm{Im}\,A(E) = \frac{k}{4\pi}\,\sigma(E)\,,$$

where $k$ is the wave vector of particles. Thus, if the total cross-section $\sigma(E)$ is known (obviously, it is distinct from zero only for $E > 0$), one can by means of the dispersion relation

$$\mathrm{Re}\,A(E) = \frac{1}{\pi}\,P\int_0^\infty \frac{dE'\,\mathrm{Im}\,A(E')}{E' - E} \tag{2.22}$$

reconstruct $\mathrm{Re}\,A(E)$. Let us note here that if there are bound states in the problem discussed, then relation (2.22) contains also a sum of contributions from the poles lying at the negative energies of these states.

Now we come back to the problem with the tuning fork. Its vibrations under an external force $f(t)$ are described by the usual oscillator equation

$$\ddot{x} + \gamma\dot{x} + \omega_0^2 x = f(t)\,, \tag{2.23}$$

where $\omega_0$ is the free oscillation frequency of the tuning fork, and $\gamma > 0$ is its damping constant. We apply to this equation the Fourier transformation (inverse with respect to (2.18)), and find

$$x(\omega) = \frac{1}{\omega_0^2 - \omega^2 - i\omega\gamma}\,f(\omega)\,. \tag{2.24}$$

Thus, the response function of a tuning fork (oscillator with friction) is

$$L(\omega) = \frac{1}{\omega_0^2 - \omega^2 - i\omega\gamma}\,. \tag{2.25}$$

Its two poles,

$$\omega_\pm = -\frac{i\gamma}{2} \pm \sqrt{\omega_0^2 - \frac{\gamma^2}{4}} = -\frac{i\gamma}{2} \pm \omega_1\,,$$

lie, indeed, in the lower half-plane $\omega$ (both for $\omega_0 > \gamma/2$, and for $\omega_0 < \gamma/2$).

Let us come back now to $L(t)$:

$$L(t) = \frac{1}{2\pi}\int_{-\infty}^\infty d\omega\, e^{-i\omega t} L(\omega)\,. \tag{2.26}$$

For $t > 0$ the integrand grows exponentially in the upper half-plane, for $\mathrm{Im}\,\omega > 0$, and decreases exponentially in the lower one, for $\mathrm{Im}\,\omega < 0$. Closing the integration contour below, where $\mathrm{Im}\,\omega \to -\infty$, we get as a result the contribution of two poles at $\omega = \omega_\pm$, which equals

$$L(t) = \frac{1}{\omega_1}\,e^{-\gamma t/2}\,\sin\omega_1 t\,. \tag{2.27}$$

On the other hand, for $t < 0$ we close the contour above, at $\mathrm{Im}\,\omega \to +\infty$, and arrive at the vanishing result. Thus, the total response function is written as follows:

$$L(t) = \frac{1}{\omega_1}\, e^{-\gamma t/2}\, \sin\omega_1 t\, \theta(t)\,, \tag{2.28}$$

where $\theta(t)$ is the step function, equal to 0 for $t < 0$, and to 1 for $t > 0$. Obviously, this is nothing but the Green function of equation (2.23).

One should emphasize how important for our conclusions is the condition $\gamma > 0$. It guarantees the stability of a tuning fork with respect to random small perturbations: oscillations caused by them die away. But for $\gamma < 0$ the system would be unstable: any small random perturbation would grow exponentially, a tuning fork would sound in fact by itself, without any regular influence $f(t)$.

### 2.4.2 Velocity of Wave Front

Let us consider now a wave packet

$$A(x,t) = \frac{1}{2\pi} \int_{-\infty}^{\infty} d\omega e^{-i\omega[t-n(\omega)x]} a(\omega)\,, \tag{2.29}$$

propagating in a medium with the refraction index $n(\omega)$ (we put the velocity of light equal to unity). This refraction index is expressed as follows via the density of atoms $N$, the electron charge $e$ and mass $m$, the so-called oscillator strength $f_{\nu 0}$ of the transition from the atomic ground state into the excited one $\nu$, the frequency $\omega_{\nu 0} = (E_\nu - E_0)/\hbar$ of this transition, and the width $\gamma_\nu$ of the excited state:

$$n(\omega) = 1 + \frac{2\pi N e^2}{m} \sum_\nu \frac{f_{\nu 0}}{\omega_{\nu 0}^2 - \omega^2 - i\omega_{\nu 0}\gamma_\nu}\,. \tag{2.30}$$

The resemblance to simple expression (2.24) is quite natural: now the atoms of matter play the part of oscillators, and $n(\omega) - 1$ is nothing but the response of medium to the incoming electromagnetic wave. As natural is the analyticity of function $n(\omega)$ in the upper half-plane.

As to the Fourier transform $a(\omega)$ of a wave, its properties by themselves are the same for a wave propagating both in a medium and in vacuum. Let us consider therefore a wave in the empty space, i.e., at $n(\omega) = 1$. We assume that at least in vacuum the discussed wave packet has a sharp front. Let at $t = 0$ the front be at the point $x = 0$. Then, obviously, the field of free wave packet

$$A_{\mathrm{free}}(x,t) = \frac{1}{2\pi} \int_{-\infty}^{\infty} d\omega e^{-i\omega(t-x)} a(\omega) \tag{2.31}$$

should vanish for $x > t$. Of course, this condition is satisfied if $a(\omega)$ is analytic in the upper half-plane. Otherwise it would be violated: when closing the integration contour above, at $\mathrm{Im}\,\omega \to +\infty$, we would obtain contributions from the singularities of the function $a(\omega)$.

Let us come back now to formula (2.29) for a wave packet in medium. Under the condition $t - x\,n(\infty) < 0$, the integration contour can be closed without any problems in the upper half-plane where the integrand decreases exponentially. Therefore, if neither $n(\omega)$ nor $a(\omega)$ have singularities in the upper half-plane, $A(x, t)$ turns to zero for $x/t > 1/n(\infty)$. It means that in a medium with the refraction index $n(\omega)$, a wave front propagates with the velocity (*M. A. Leontovich*, 1937) [5]

$$v_f = \frac{1}{n(\infty)}. \tag{2.32}$$

Thus, in optics, where $n(\infty) = 1$ (see formula (2.30)), the wave front propagates with the velocity of light. This fact is well-known by experimentalists: the first signal, so-called precursor, reaches a detector earlier than the bulk of a wave packet, the center of which moves with the group velocity.

In fact, there is nothing surprising about it. The wave packet with a sharp front contains high frequencies. And for them the refraction index (2.30) tends to unity. Just these high-frequency components form the precursor.

---

[5]The presentation here follows a paper by *A. D. Dolgov, I. B. Khriplovich* (1981).

# 3

# Atomic Physics. Minimum Calculations

## 3.1 Semiclassical Hydrogen Atom

### 3.1.1 Are Circular Orbits Semiclassical?

Let us consider an electron state in a hydrogen atom characterized by quantum numbers $n \gg 1$ and $l = n - 1$, which corresponds, as it is known, to circular orbits. Is the radial wave function $R_{n,n-1}(r)$ of this state semiclassical? At first sight, no. Indeed, the radial quantum number $n_r = n - l - 1$ is in this case in no way large, but is equal to zero, so that the wave function $R_{n,n-1}(r)$ has no nodes at all.

However, let us look at the problem more attentively. Compare to this end the radial distance at which the probability density

$$r^2 R_{n,n-1}^2(r) \propto r^{2n} e^{-2r/n} \tag{3.1}$$

is maximal, with the width of this maximum (here we put the Bohr radius equal to unity, and omit the normalization factor, inessential in the present case). The density (3.1) is maximal at $r_0 = n^2$. To estimate its width, we rewrite expression (3.1) in the exponential form, and expand the exponent into series near the maximum up to second order included:

$$r^2 R_{n,n-1}^2(r) \propto \exp(-2r/n + 2n \ln r) \approx$$
$$\approx \exp[-2n + 4n \ln n - (r - r_0)^2/n] \, . \tag{3.2}$$

It is clear from expression (3.2) that the width of the probability distribution in $\Delta r = |r - r_0|$ equals $\sqrt{n}$, and thus is small as compared with the position $r_0 = n^2$ of the maximum.

Therefore, from this point of view, which is the most natural one, the discussed state is certainly semiclassical. In other words, for a state to be semiclassical, it is not required at all that its radial quantum number $n_r$ is large. Of course, in line with the state where $n \gg 1$, $l = n - 1$, and $n_r = 0$, all the states with $n \gg 1$, $l \gg 1$ are semiclassical as well.

### 3.1.2 Approximate Selection Rule for Electromagnetic Transitions

It is well-known that in the dipole transitions dominating in atoms, the selection rule for the orbital angular momentum is $\Delta l = \pm 1$. Meanwhile, the classical radiation is always accompanied by the loss of angular momentum. Thus, at least in the semiclassical limit the probability of dipole transitions with $\Delta l = -1$ is higher. Here we discuss the question how strongly and under what conditions the transitions with $\Delta l = -1$ are dominating in atoms. Our approach to the problem is based on classical electrodynamics and, of course, on the correspondence principle [1]. The obtained results describe not only the semiclassical situation. In a remarkable way, they agree, at least qualitatively, with exact relations which refer to absolutely nonclassical transitions with $|\Delta n| \sim n \sim 1$ and $l \sim 1$ [2]. (To simplify the presentation, we mean always, here and below, the radiation of a photon, i.e., transitions with $\Delta n < 0$. Obviously, in the case of photon absorption, i.e., for $\Delta n > 0$, the angular momentum predominantly increases.)

We start our analysis with a purely classical problem. Let a classical particle with mass $m$ and charge $e$ move in an attractive Coulomb field along an ellipse with large semi-axis $a$ and eccentricity $\varepsilon$. As is known [3], the radiation intensity at a given harmonic $\nu$ is here

$$I_\nu = \frac{4e^2\omega_0^4\nu^4 a^2}{3c^3}\left(\xi_\nu^2 + \eta_\nu^2\right);\tag{3.3}$$

$$\xi_\nu = \frac{1}{\nu}J_\nu'(\nu\varepsilon),\quad \eta_\nu = \frac{\sqrt{1-\varepsilon^2}}{\nu\varepsilon}J_\nu(\nu\varepsilon);\quad \nu > 0.\tag{3.4}$$

In expressions (3.4) $J_\nu(\nu\epsilon)$ is the Bessel function, and $J_\nu'(\nu\epsilon)$ is its derivative. We note that since $0 \leq \epsilon \leq 1$, both $J_\nu(\nu\epsilon)$ and $J_\nu'(\nu\epsilon)$ are reasonably well approximated by the first term of their series expansion in the argument. Then it becomes clear that all the Fourier components $\xi_\nu$ and $\eta_\nu$ are positive.

It is convenient to use the Fourier transformation in the following form:

$$x(t) = a\sum_{\nu=-\infty}^{\infty}\xi_\nu\, e^{i\nu\omega_0 t} = 2a\sum_{\nu=0}^{\infty}\xi_\nu\cos\nu\omega_0 t,$$

$$y(t) = -i\,a\sum_{\nu=-\infty}^{\infty}\eta_\nu\, e^{i\nu\omega_0 t} = 2a\sum_{\nu=0}^{\infty}\eta_\nu\sin\nu\omega_0 t.$$

---

[1] We follow in this section the work by *I.B. Khriplovich, D.V. Matvienko* (2007).

[2] The analysis of numerical values of transition probabilities (*H. Bethe, E. Salpeter*, 1957) has demonstrated that even for $n$ and $l$ comparable with unity, i.e., in a nonclassical situation, radiation with $\Delta l = -1$ is considerably more probable than radiation with $\Delta l = 1$.

[3] See: L.D. Landau and E.M. Lifshitz, *The Classical Theory of Fields*. §70.

Since $x(t)$, $y(t)$ are real, all dimensionless Fourier components $\xi_\nu$ and $\eta_\nu$ are real as well, and $\xi_{-\nu} = \xi_\nu$, $\eta_{-\nu} = -\eta_\nu$. We note that the Cartesian coordinates $x$ and $y$ are related here to the polar coordinates $r$ and $\phi$ as follows: $x = r\cos\phi$, $y = r\sin\phi$, where $\phi$ increases with time. Thus, the angular momentum is directed along the $z$ axis (but not in the opposite direction).

In the quantum problem, where $\nu = |\Delta n|$, the probability of transition in unit time is

$$W_\nu = \frac{I_\nu}{\hbar\omega_0\nu} = \frac{4e^2\omega_0^3\nu^3 a^2}{3c^3\hbar}\left(\xi_\nu^2 + \eta_\nu^2\right), \quad \omega_0 = \frac{me^4}{\hbar^2 n^3}. \tag{3.5}$$

Now, the loss of angular momentum with radiation is [4]

$$\dot{\mathbf{M}} = \frac{2e^2}{3c^3}\,\mathbf{r} \times \dddot{\mathbf{r}}.$$

Going over here to the Fourier components, we obtain

$$\dot{\mathbf{M}}_\nu = -\frac{4e^2\omega_0^2\nu^2}{3c^3}\,\mathbf{r}_\nu \times \dot{\mathbf{r}}_\nu,$$

or (with our choice of the direction of coordinate axes, and with the angular momentum measured in the units of $\hbar$)

$$\dot{M}_\nu = -\frac{4e^2\omega_0^3\nu^3 a^2}{3c^3\hbar}\,2\xi_\nu\eta_\nu. \tag{3.6}$$

Obviously, the last expression is nothing but the difference between the probabilities of transitions in the unit time with $\Delta l = 1$ and $\Delta l = -1$:

$$\dot{M}_\nu = W_\nu^+ - W_\nu^-. \tag{3.7}$$

Of course, the total probability (3.5) can be written as

$$W_\nu = W_\nu^+ + W_\nu^-. \tag{3.8}$$

From explicit expressions (3.5) and (3.6) it is clear that inequality $W_\nu^+ \ll W_\nu^-$ holds if $2\xi_\nu\eta_\nu \approx \xi_\nu^2 + \eta_\nu^2$, or $\eta_\nu \approx \xi_\nu$. The last relation is valid for $\epsilon \ll 1$, i.e., for orbits close to circular ones. (The simplest way to check it, is to use in formulas (3.4) the explicit expression for the Bessel function at small argument: $J_\nu(\nu\epsilon) = (\nu\epsilon)^\nu/(2^\nu\nu!)$.)

This conclusion looks quite natural from the quantum point of view. Indeed, it is the state with the orbital quantum number $l$ equal to $n-1$ (i.e., with the maximum possible value for given $n$) which corresponds to the circular orbit. As a result of radiation $n$ decreases, and therefore $l$ should decrease as well.

---

[4]See: L.D. Landau and E.M. Lifshitz, *The Classical Theory of Fields*. §75.

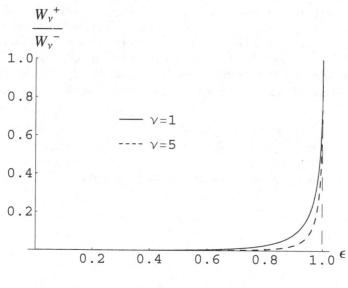

$$\frac{W_\nu{}^+}{W_\nu{}^-}$$

——— $\nu=1$

---- $\nu=5$

*Fig. 3.1*

The surprising fact is, however, that the probabilities $W_\nu^-$ of transitions with $\Delta l = -1$, in fact, dominate numerically everywhere, except for a small vicinity of the maximum possible eccentricity $\varepsilon = 1$. For instance, if $\varepsilon \simeq 0.9$ (which is much closer to 1 than to 0 !), then at $\nu = 1$ the discussed probability ratio is very large, it constitutes

$$\frac{W_\nu^-}{W_\nu^+} \simeq 12 \,.$$

The dependence on $\varepsilon$ of the ratio of $W_\nu^+$ to $W_\nu^-$ for two different values of $\nu$ is presented in Fig. 3.1. With increase of $\nu$, the region where $W_\nu^-$ and $W_\nu^+$ are comparable gets more and more narrow, i.e., the corresponding curves tend more and more to the right angle. For $\nu \gtrsim 1$ it follows from direct numerical calculations.

For $\nu \gg 1$ it can be demonstrated analytically. To this end we present the discussed ratio as follows:

$$\frac{W_\nu^+}{W_\nu^-} = \left( \frac{1 - \eta_\nu/\xi_\nu}{1 + \eta_\nu/\xi_\nu} \right)^2 \,.$$

In the situation of interest to us, where $\nu \gg 1$ and $1 - \varepsilon \ll 1$, the asymptotic relation holds:

$$J_\nu(\nu\varepsilon) = \frac{1}{\sqrt{\pi}} \left( \frac{2}{\nu} \right)^{1/3} \Phi \left[ \left( \frac{\nu}{2} \right)^{2/3} (1 - \varepsilon^2) \right], \qquad (3.9)$$

where $\Phi$ is the Airy function [5]. The ratio $-\eta_\nu/\xi_\nu$ is here

$$-\frac{\eta_\nu}{\xi_\nu} = x^{1/2}\frac{\Phi(x)}{\Phi'(x)}, \quad x = \left(\frac{\nu}{2}\right)^{2/3}(1-\varepsilon^2). \tag{3.10}$$

For $x \simeq 1$ ratio (3.10) can be estimated (with $\Phi(0) = 0.629$, $\Phi'(0) = -0.459$) as

$$-\frac{\eta_\nu}{\xi_\nu} \simeq x^{1/2}\frac{\Phi(0)}{\Phi'(0)} \simeq -1.37\,\nu^{1/3}(1-\varepsilon)^{1/2}. \tag{3.11}$$

Now, the ratio of probabilities $W_\nu^+/W_\nu^-$ is close to unity when $\eta_\nu/\xi_\nu$ is small. Obviously, for this ratio to be small, $\varepsilon$ should be ever closer to unity with the increase of $\nu$.

For $x \gg 1$ the Airy function behaves as follows:

$$\Phi(x) \simeq \frac{1}{2x^{1/4}}\exp\left(-\frac{2}{3}x^{3/2}\right),$$

so that $-\eta_\nu/\xi_\nu \to -1$, and the ratio $W_\nu^+/W_\nu^-$ vanishes. However, this region, $x \gg 1$, is of no real interest since even the probability $W_\nu^-$ is here exponentially small.

Let us go over now to the quantum problem. In the semiclassical limit, classical expression for the eccentricity

$$\varepsilon = \sqrt{1 + \frac{2EM^2}{me^4}} \tag{3.12}$$

is rewritten with usual relations $E = -me^4/(2\hbar^2 n^2)$ and $M = \hbar l$ as

$$\varepsilon = \sqrt{1 - \frac{l^2}{n^2}}. \tag{3.13}$$

We note that the exact expression for $\varepsilon$, valid for arbitrary $l$ and $n$, is [6]

$$\varepsilon = \sqrt{1 - \frac{l(l+1)+1}{n^2}}. \tag{3.14}$$

Clearly, in the semiclassical approximation, the eccentricity is close to unity only under condition $l \ll n$. If this condition does not hold, one may expect that in the semiclassical limit the transitions with $\Delta l = -1$ dominate. In other words, as long as $l \ll n$, the probabilities of transitions with decrease and increase of the angular momentum are comparable. But if the angular momentum is not small, it is being lost predominantly in radiation. This situation looks quite natural.

The analysis of the numerical values of transition probabilities (*H. Bethe, E. Salpeter*, 1957) demonstrates that even for $n$ and $l$ comparable with unity

---

[5]See: L.D. Landau and E.M. Lifshitz, *The Classical Theory of Fields*. §70.
[6]See: L.D. Landau and E.M. Lifshitz, *Quantum Mechanics*. §36.

Table 3.1

| | $\dfrac{W_{4p\to3s}}{W_{4p\to3d}}$ | $\dfrac{W_{5p\to4s}}{W_{5p\to4d}}$ | $\dfrac{W_{5d\to4p}}{W_{5d\to4f}}$ | $\dfrac{W_{6f\to5d}}{W_{6f\to5g}}$ | $\dfrac{W_{5p\to3s}}{W_{5p\to3d}}$ | $\dfrac{W_{6p\to3s}}{W_{6p\to3d}}$ |
|---|---|---|---|---|---|---|
| exact value | 10 | 3.75 | 28 | 72 | 10.67 | 13.7 |
| $\bar{\varepsilon}$ | 0.87 | 0.92 | 0.81 | 0.75 | 0.90 | 0.92 |
| $\nu = \Delta n$ | 1 | 1 | 1 | 1 | 2 | 3 |
| semiclassical value | 17.6 | 8.7 | 34 | 58 | 17.2 | 15.7 |

and $\Delta n \simeq n$, i.e., in the absolutely nonclassical region, the transitions with $\Delta l = -1$ still are much more probable than those with $\Delta l = 1$. The results of this analysis for the ratio $W^-/W^+$ in some transitions are presented in Table 3.1 (first line). Then we indicate in Table 3.1 the values of these ratios obtained in the naïve semiclassical approximation. Here for the eccentricity $\bar{\varepsilon}$ we use the value of expression (3.14), calculated with $l$ corresponding to the initial state; as to $n$, we take its average value for the initial and final states. The table demonstrates that the ratio of the naïve semiclassical results to the exact ones remains within a factor of about two. Moreover, if one uses as $\bar{\varepsilon}$ expression (3.13), calculated in the analogous way, the numbers in the last line change considerably. It is clear, nevertheless, that the semiclassical approximation describes here, at least qualitatively, the real problem.

### 3.1.3 Expectation Value $\langle r \rangle$ in Hydrogen Atom, and Lattice Constant for Alkali Metals

Let us come back to a classical particle with mass $m$, moving in the attracting Coulomb field $-e^2/r$ along an ellipse with eccentricity $\varepsilon$, and find the mean value of its radius $< r >$. The equation of motion of the particle is

$$r = \frac{p}{1 + \varepsilon \cos \phi} ; \tag{3.15}$$

here $p = L^2/(me^2)$, where $L$ is the classical angular momentum of the particle (compare with (1.17)). The discussed mean value is

$$<r> = \frac{1}{T} \int_0^T dt\, r(t)\,; \tag{3.16}$$

here $T = \pi e^2 \sqrt{m/(2|E|^3)}$ is the period of particle rotation. Let us go over in the last expression from integration over time to integration over angle $\phi$ with the relation $L = mr^2 d\phi/dt$ for the angular momentum:

$$<r> = \frac{m}{TL} \int_0^{2\pi} d\phi\, r^3(\phi) = \frac{mp^3}{TL} \int_0^{2\pi} \frac{d\phi}{(1 + \varepsilon \cos \phi)^3}\,. \tag{3.17}$$

The factor at the last integral transforms to

$$\frac{mp^3}{TL} = \frac{L^5}{me^2} \left(\frac{2|E|}{me^4}\right)^{3/2} \frac{1}{2\pi}\,,$$

and the integral itself is

$$\int_0^{2\pi} \frac{d\phi}{(1 + \varepsilon \cos \phi)^3} = \frac{\pi(2 + \varepsilon^2)}{(1 - \varepsilon^2)^{5/2}}\,.$$

Finally, after simple transformation with the account for relation (3.12), we find

$$<r> = \frac{1}{me^2} \frac{1}{2} \left(\frac{3me^4}{2|E|} - L^2\right). \tag{3.18}$$

For the hydrogen atom in the semiclassical limit, this expression can be rewritten with the usual relations $E = -me^4/(2\hbar^2 n^2)$ and $M = \hbar l$, as

$$<r> = \frac{\hbar^2}{me^2} \frac{1}{2} \left(3n^2 - l^2\right).$$

It is sufficient to change here, as usual, $l^2 \to l(l+1)$, and we arrive at the exact quantum-mechanical expression for the expectation value under discussion:

$$\langle nl|r|nl\rangle = \frac{\hbar^2}{me^2} \frac{1}{2} \left[3n^2 - l(l+1)\right] = a \frac{1}{2} \left[3n^2 - l(l+1)\right]\,; \tag{3.19}$$

here $a$ is the Bohr radius.

Let us go over now from hydrogen to alkali atoms. In the ground state of such atoms, the outer electron is in the $ns_{1/2}$ state. If this state is semiclassical, its energy can be found with the Bohr–Sommerfeld quantization rule for the radial motion. At large distances from the nucleus, the outer electron moves in the Coulomb field with charge $Z = 1$. However, at smaller distances the field is rather different. This deviation from the Coulomb field can be described formally as a change of the boundary condition at $r = 0$ for the wave function of the outer electron. This change results in the change of the constant in the quantization rule: $n_r$ goes over into $n_r - \sigma$, and correspondingly, $n$ goes over into $n^* = n - \sigma$. Thus, the energy of the outer electron in an alkali atom is

Table 3.2

|  | Li | Na | K | Rb | Cs |
|---|---|---|---|---|---|
| $2 < r >= 3\mathrm{Ry}/\mathrm{I}$ | 7.57 | 7.94 | 9.41 | 9.77 | 10.48 |
| $d/2$ | 5.72 | 7.00 | 8.72 | 9.19 | 9.89 |
| $\bar{a}$ | 6.61 | 8.08 | 10.06 | 10.60 | 11.42 |

$$E_n = -\frac{me^2}{2\hbar^2 (n - \sigma)^2} = -\frac{me^2}{2\hbar^2 n^{*2}}.$$

Now, due to relation (3.18), the mean orbit radius for the outer $s$ electron is (in the units of the Bohr radius)

$$< r >= \frac{3}{2} n^{*2}. \tag{3.20}$$

It is natural to expect that the following estimate will be correct at least qualitatively for the lattice constant of an alkali metal. This constant is about twice as large as the mean radius of the free atom, i.e., it is close to

$$2 < r >= 3 \frac{\mathrm{Ry}}{\mathrm{I}}; \tag{3.21}$$

here $\mathrm{Ry} = me^4/(2\hbar^2) = 13.6$ eV is the Rydberg constant, and I is the ionization potential of the free atom.

Let us compare this prediction with the real values of the lattice constants. All alkalis have body-centered cubic cells. Besides the edge of the cube $\bar{a}$, there is here even shorter interatomic distance, which is half of the main diagonal of the cube $d/2 = \bar{a}\sqrt{3}/2$. The results of comparison are presented in Table 3.2 (all values therein are given in units of the Bohr radius). Even for lithium, where $n = 2$ and one could hardly expect that our semiclassical estimate is applicable, its error constitutes about 25% only. For all other metals our estimate hits the interval between $d/2$ and $\bar{a}$, i.e., it proves to be correct quantitatively.

To paraphrase the known saying on mathematics and natural sciences, this problem can be called an example of "inconceivable efficiency" of semiclassical approximation in quantum mechanics.

## 3.2 Simple Calculation of Expectation Values $< 1/r^n >$ in Hydrogen Atom

The simplest calculation is that of the matrix element $\langle 1/r \rangle$. Here it is sufficient to recall the virial theorem, according to which the mean value of the potential energy $- e^2/r$ of a particle bound by a Coulomb field is equal to its doubled total energy, i.e., this mean value is $- me^4/(\hbar^2 n^2)$. It follows now immediately that

$$\left\langle \frac{1}{r} \right\rangle = \frac{me^2}{\hbar^2 n^2} = \frac{1}{a\,n^2}, \tag{3.22}$$

where $a$ is the Bohr radius.

The next matrix element is $\langle 1/r^2 \rangle$. Here it is convenient to use the fact that the derivative of energy $E$ over a parameter is equal to the mean value of the derivative of Hamiltonian $H$ over this parameter:

$$\frac{\partial E_n}{\partial \lambda} = \left\langle n \left| \frac{\partial \hat{H}}{\partial \lambda} \right| n \right\rangle. \tag{3.23}$$

Therefore, for an arbitrary central field with the Hamiltonian of radial motion

$$\hat{H}_r = -\frac{\hbar^2}{2m} \left( \frac{\partial^2}{\partial r^2} + \frac{2}{r} \frac{\partial}{\partial r} \right) + \frac{\hbar^2 l(l+1)}{2mr^2} + U(r) \tag{3.24}$$

we obtain

$$\frac{\partial E_{n_r l}}{\partial l} = \left\langle n_r l \left| \frac{\partial H_r}{\partial l} \right| n_r l \right\rangle = \left\langle n_r l \left| \frac{\hbar^2 (2l+1)}{2mr^2} \right| n_r l \right\rangle. \tag{3.25}$$

In particular, for the hydrogen atom, where

$$E_{n_r l} = -\frac{me^4}{2\hbar^2 (n_r + l + 1)^2},$$

we find easily from (3.25) that

$$\left\langle nl \left| \frac{1}{r^2} \right| nl \right\rangle = \frac{m^2 e^4}{\hbar^4 n^3 (l + \frac{1}{2})} = \frac{1}{a^2 n^3 (l + \frac{1}{2})}. \tag{3.26}$$

Let us note that in the limit of large quantum numbers their total power in both expectation values, (3.22) and (3.26), as well as in formula (3.19) for $< r >$, coincides with the power of $\hbar$. Of course, it should be so for any matrix element which has a classical limit.

And now we go over to the matrix elements $\langle \delta(\mathbf{r}) \rangle = |\psi(0)|^2$ and $\langle 1/r^3 \rangle$. Here we use the identity

$$\langle nl | \left[ \frac{d}{dr}, H_r \right] | nl \rangle = 0,$$

where the Hamiltonian of radial motion $H_r$ is given by formula (3.24) with $U(r) = -e^2/r$. The explicit calculation of the commutator gives

$$\left[\frac{d}{dr}, H_r\right] = \frac{1}{mr^2}\frac{d}{dr} - \frac{l(l+1)}{mr^3} + \frac{e^2}{r^2}.$$

The value of $\langle 1/r^2\rangle$ was found earlier (see (3.26)). The first term on the right-hand side transforms as follows:

$$\left\langle \frac{1}{r^2}\frac{d}{dr}\right\rangle = \int d\Omega \int_0^\infty \psi^* \frac{1}{r^2}\frac{d\psi}{dr} r^2\, dr$$

$$= \frac{1}{2}\int d\Omega \int_0^\infty \frac{d|\psi|^2}{dr}\, dr = -2\pi|\psi(0)|^2.$$

It differs from zero for $l = 0$ only. Thus, we find

$$\langle nl|\,\delta(\mathbf{r})\,|nl\rangle = |\psi_{nl}(0)|^2 = \frac{m^3 e^6}{\pi\hbar^6 n^3}\delta_{l0} = \frac{1}{\pi a^3 n^3}\delta_{l0}. \tag{3.27}$$

And for $l \neq 0$ we obtain

$$\left\langle nl\left|\frac{1}{r^3}\right|nl\right\rangle = \frac{m^3 e^6}{\hbar^6 n^3 l(l+1)(l+1/2)}\,(1 - \delta_{l0})$$

$$= \frac{1}{a^3 n^3 l(l+1)(l+1/2)}\,(1 - \delta_{l0}). \tag{3.28}$$

It is noteworthy that in the expectation value $\langle 1/r^3\rangle$ as well, the power $-6$ of $\hbar$ coincides in the classical limit $l \gg 1$ with the total power of quantum numbers in the product $n^{-3}l^{-3}$. It is not the case, however, in the matrix element (3.27), and this is quite natural: in the discussed problem the probability density at the origin, $|\psi_{nl}(0)|^2$, vanishes in the classical limit.

## 3.3 Thomas–Fermi Method

In the Thomas–Fermi approximation, the main qualitative characteristics of a many-electron atom can be obtained with simple and intuitive considerations.

With the increase of the nuclear charge $Z$, the radius and volume of a neutral atom do not change essentially. Both the radius of an atom and its volume are determined by the external electron, which moves in the Coulomb field of the nucleus screened by internal electrons. Due to this screening, the external electron is mainly in an electric field of the same order of magnitude as that in the hydrogen atom. However, since in the case of a many-electron atom in its volume $\sim a^3$ there are $Z$ electrons, the mean distances both from an electron to the nucleus and between electrons are estimated as $\sim a/Z^{1/3}$.

Somewhat more detailed derivation looks as follows. Let us write the density of electrons as [7]

$$n(r) = Z^\alpha F(rZ^\beta) \,. \tag{3.29}$$

Then, their total number is

$$Z = 4\pi \int_0^\infty dr\, r^2\, n(r) = 4\pi Z^{\alpha - 3\beta} \int_0^\infty dx\, x^2 F(x) \,,$$

so that $\alpha - 3\beta = 1$.

In virtue of (3.29), the mean distance from an electron to the nucleus is

$$\langle r \rangle = \frac{\int_0^\infty dr\, r^3\, n(r)}{\int_0^\infty dr\, r^2\, n(r)} \sim Z^{-\beta} \,.$$

On the other hand, this distance behaves as $Z^{-1/3}$. Hence $\beta = 1/3$. Now, in virtue of $\alpha - 3\beta = 1$, we get $\alpha = 2$. As a result, the electron density is

$$n(r) = Z^2\, F(rZ^{1/3}). \tag{3.30}$$

Let us consider now the electrostatic potential created by an atom. This potential changes from $Z/r$ at $r \to 0$, where the nucleus is not screened by electrons, to zero at $r \to \infty$, where the nucleus is screened completely. Since the characteristic distance, at which the density of electrons changes, behaves as $Z^{-1/3}$, then the potential is naturally written as

$$\phi(r) = \frac{Z}{r}\, f(rZ^{1/3}) \,, \tag{3.31}$$

where $f(0) = 1$, and $f(\infty) = 0$. The mean potential energy of an electron is $-\phi(r)$.

Due to the virial theorem, the mean kinetic energy $T$ of an electron and its mean potential energy $U$ are comparable in modulus. Therefore, the estimate for the mean electron momentum is

$$\langle p \rangle \sim \sqrt{T} \sim \sqrt{|U|} \sim \sqrt{Z/\langle r \rangle} \sim Z^{2/3} \,. \tag{3.32}$$

In this way we arrive at the estimate for the mean electron angular momentum:

$$\langle l \rangle \sim \frac{\langle r \rangle \cdot \langle p \rangle}{\hbar} \sim Z^{1/3}. \tag{3.33}$$

Let us use relation (3.33) between $l$ and $Z$ to estimate the critical value $Z_l$, at which electrons first appear in the shell with given $l$. To this end, we just put

$$Z_l \sim l^3 \,. \tag{3.34}$$

---

[7]From now on in this section we use atomic units.

This estimate proves to be quite reasonable. Indeed, a numerical calculation in the Thomas–Fermi approximation gives

$$Z_l = 1.24 \, (l + 1/2)^3 \, . \tag{3.35}$$

The transition in the semiclassical approximation from $l$ to $l + 1/2$ is quite natural, and the numerical factor here is close to unity. The empirical relation is

$$Z_l = 1.36 \, (l + 1/2)^3 \, . \tag{3.36}$$

The total energy of an atom $E$ is about $Z$ times larger than the mean potential energy of an electron, i.e.,

$$E \sim -Z \cdot Z \cdot (Z^{-1/3})^{-1} = -Z^{7/3}; \tag{3.37}$$

Here the first and second factors $Z$ are the number of electrons and the charge of the nucleus, respectively, and the factor $(Z^{-1/3})^{-1}$ is the inverse mean distance of an electron from the nucleus. Numerical calculations in the Thomas–Fermi approximation result in $E = -0.76 \, Z^{7/3}$ (in atomic units, i.e., in the units of $me^4/\hbar^2 = 2\,\mathrm{Ry} = 27.2\,\mathrm{eV}$). Empirically, this dependence of the total energy on $Z$ is confirmed, though with a factor close to 0.59. Again, simple estimate (3.37) works reasonably well.

The used semiclassical analysis is certainly inapplicable to the first Bohr orbit. Its radius is $\sim Z^{-1}$, and the nuclear charge $Z$ is not screened here. So, the energy of these two electrons is on the order of $\sim Z^2$. This is the leading correction to the total energy (3.37), its relative magnitude is $\sim Z^{-1/3}$.

## 3.4 *LS*-Coupling Approximation. Explanation of the Second Hund's Rule

In many-electron atoms one can, with fair accuracy, describe the state of each electron as a stationary state in some effective centrally symmetric field created by the nucleus and by all the other electrons. Deviations from this approximation in light atoms are due mainly to that residual part of the Coulomb interaction, which has no spherical symmetry and therefore cannot be reduced to a central potential. The energy splitting between atomic levels having different total orbital angular momenta $L$ and total spin $S$, is described in light atoms by Hund's rules.

According to the first Hund's rule, the state with the maximum possible total electron spin $S$ for the given electron configuration has the minimum energy. The qualitative explanation of this rule is well-known. In the state with the maximum total electron spin $S$, the spin wave function is "most symmetrical". Therefore, by virtue of the Pauli principle, the coordinate wave function of this state is "most antisymmetrical". This minimizes the probability of finding electrons close to each other, and thus minimizes the electron–electron Coulomb repulsion.

But what is the origin of the second Hund's rule according to which, for a given $S$, it is the state with the maximum possible total orbital angular momentum $L$ that has the lowest energy? In some textbooks this rule, first established empirically, is confirmed by direct (and rather tedious) calculations of the residual electron–electron Coulomb interaction. Here we present a simple and transparent explanation of this rule.

Without any claims of generality, we confine ourselves to a simple example of the system of two equivalent $p$-electrons, $p^2$. According to the first Hund's rule, its lowest terms are the triplet ones, $^3P_{0,1,2}$.

As to the singlet terms, $^1S_0$ and $^1D_2$, their radial wave functions are the same if one neglects the residual electron–electron interaction. Let us compare, however, their angular wave functions. For a single $p$-electron this wave function is nothing but a normalized component, Cartesian or spherical one, of the unit radius-vector, $\sqrt{3}\,n_i$. Then the total $^1S_0$ angular wave function, again properly normalized, is

$$\sqrt{3}\,(\mathbf{n}^{(1)} \cdot \mathbf{n}^{(2)}),\tag{3.38}$$

where the superscripts mark the unit vectors of the first and second electron, respectively. As one of the $^1D_2$ states, let us choose, for instance, that with $L_z = +2$. Its angular wave function is just the product of the single-electron wave functions corresponding to $l_z = +1$ each:

$$\sqrt{\frac{3}{2}}\,(-\mathbf{n}_x^{(1)} - i\mathbf{n}_y^{(1)}) \cdot \sqrt{\frac{3}{2}}\,(-\mathbf{n}_x^{(2)} - i\mathbf{n}_y^{(2)}).\tag{3.39}$$

Let us consider the extreme case of the coinciding coordinates of the two electrons. For $\mathbf{n}^{(1)} = \mathbf{n}^{(2)}$, wave function (3.38) squared equals 3. And the modulus squared of wave function (3.39) is

$$\frac{9}{4}\,(n_x^2 + n_y^2)^2.$$

Even its maximum value, $9/4$, is less than 3.

It is obvious from this example that the probability of finding the two electrons close to each other in the $D$-state is less than in the $S$-state. Therefore, the Coulomb repulsion in the $D$-state is more weak, and the $D$-state energy is, correspondingly, lower.

Let us note that the wave function (3.38) is a scalar, it is invariant under arbitrary rotations in the three-dimensional space. As to the wave function (3.39), its symmetry is lower: it is invariant only under rotations around the $z$ axis, and only by modulus.

We note also that the $^1S_0$ state, which is of higher symmetry, generally speaking is unstable. In principle, there is a radiative electric quadrupole transition from it into the less symmetric state $^1D_2$. Of course, the probability of this transition is tiny [8]. Nevertheless, there is an undoubted resemblance here to the well-known Jahn–Teller rule according to which in a degenerate electron state of a molecule a symmetrical configuration of the nuclei is unstable.

To emphasize once more how essential is for our conclusions the fact that the residual electron–electron interaction is repulsive, we consider the following example. Let electrons move in an attractive Coulomb (or Newton) field, and the residual interaction between them is not repulsive, but attractive (gravitational atom). Obviously, in such a system the second Hund's rule (and the first one as well) changes to the opposite one.

In conclusion of this section, we mention the third Hund's rule: for given $L$ and $S$, the energy of a shell filled less than by half grows with the total angular momentum $J$. This rule originates from the spin–orbit interaction due to which the energy of an electron grows with its total angular momentum $j$. However, for a hole the sign of the spin–orbit interaction is opposite. Therefore, if a shell is filled more than by half, its energy decreases with the growth of $J$. As to a shell filled exactly by half, in it the splitting in $J$ is absent to first order in the spin–orbit interaction.

## 3.5 $jj$-Coupling Approximation. Intermediate Coupling

Let us consider now the opposite limiting case, that of the $jj$-coupling, when the spin–orbit interaction is essentially stronger than the residual Coulomb one. When neglecting that Coulomb interaction, the atomic Hamiltonian corresponds to a set of noninteracting electrons, each of which moves in the potential

$$V_{\text{eff}}(r_a) + A(r_a)\,\hat{\mathbf{l}}_a\hat{\mathbf{s}}_a \ .$$

Since both $V_{\text{eff}}(r_a)$ and $A(r_a)$ are centrally symmetric, in such a field both the total angular momentum $j = l \pm 1/2$ of a single electron and its projection $m_j$ are conserved. From the single-electron states $|\,nljm_j\rangle$ one constructs (taking into account the Pauli principle) the atomic states with definite $J$ and $M_J$. Then, already with these states, one finds the corrections to the energy of an atom caused by a residual interaction.

In fact, usually the case of $jj$-coupling in its pure form does not occur in atoms. Though the spin–orbit interaction grows with $Z$, even in heavy atoms it only reaches the same order of magnitude as the residual Coulomb one. The simultaneous account of both interactions, the so-called intermediate coupling approximation, requires usually sufficiently tedious calculations. We will demonstrate, however, that even here important qualitative results can be derived in a sufficiently simple way.

---

[8] We note that the magnetic dipole transition $^1S_0 \rightarrow\ ^3P_1$ is forbidden. As is known, the magnetic moment operator is proportional to $L_z + 2S_z$. Obviously, neither $L_z$ nor $S_z$ can transform $^1S_0$ into $^3P$, as well as $^1D_2$ into $^3P$. Thus, in the given electron configuration $p^2$ only the following $M1$ transitions are possible: $^3P_2 \rightarrow\ ^3P_1 \quad ^3P_1 \rightarrow\ ^3P_0$.

### 3.5.1 Lead, Configuration $p^2$

We start with the limiting case of $jj$-coupling when an electron is characterized only by its total angular momentum $j$, equal for $p$-electron to $1/2$ or $3/2$. A state of two $p$-electrons is described here by a set $(j_1 j_2)_J$ where the total angular momentum runs through the values $J = 0, 1, 2$. Then the possible states are:

$$(1/2\ 1/2)_0\,;\quad (1/2\ 3/2)_1\,,\quad (1/2\ 3/2)_2\,;\quad (3/2\ 3/2)_0\,,\quad (3/2\ 3/2)_2\,. \quad (3.40)$$

The state $(1/2\ 1/2)_0$ is the only one possible for $j_1 = j_2 = 1/2$. Indeed, by virtue of the Pauli principle, one cannot construct here a state with $J_z = \pm 1$. Hence, $J = 1$ is impossible here as well. By an analogous reason (one cannot construct $J_z = \pm 3$), the states $j_1 = 3/2$, $j_2 = 3/2$ do not add up into $J = 3$. The state $(3/2\ 3/2)$ with $J_z = +1$ can be constructed in only one way: from the single-electron projections $-1/2$ and $+3/2$. However, one such state should belong to $J = 2$, so that the state $(3/2\ 3/2)_1$ does not exist. Since an electron with larger $j$ has larger energy, the sequence of levels, ordered in accordance with the energy increase, looks as presented in (3.40), where commas and semicolons separate states of the same energies and different energies, respectively.

Now we recall that in the approximation of $LS$-coupling, according to Hund's rules, the sequence of levels, in the order of energy increase, should be as follows:

$$^3P_0;\quad ^3P_1;\quad ^3P_2;\quad ^1D_2;\quad ^1S_0. \quad (3.41)$$

As should be expected, the number of states with given total angular momentum $J$ is the same both in the $LS$- and $jj$-schemes of addition for angular momenta. Besides, it follows from the comparison of schemes (3.40) and (3.41) that the states $(1/2\ 3/2)_1$ and $^3P_1$ coincide. At last, the same comparison demonstrates that in the case intermediate between the two limiting ones, $jj$ and $LS$, the states are ordered as follows:

$$J = 0;\quad J = 1;\quad J = 2;\quad J = 2;\quad J = 0.$$

Here the states with $J = 0$ are orthogonal linear combinations of $^3P_0$ and $^1S_0$ in the $LS$-scheme, or $(1/2\ 1/2)_0$ and $(3/2\ 3/2)_0$ in the $jj$ one. Analogously, the states with $J = 2$ are orthogonal linear combinations of $^3P_2$ and $^1D_2$, or $(1/2\ 3/2)_2$ and $(3/2\ 3/2)_2$.

### 3.5.2 Bismuth, Configuration $p^3$

We start the analysis of this configuration with the $LS$-coupling approximation. In the state with the maximum total spin $S = 3/2$, the spin wave function of three electrons is completely symmetric (this is quite obvious for its components with $S = \pm 3/2$, where the projections of all three spins should be the same). But then the coordinate wave function, in accordance with the Pauli principle, is totally antisymmetric. For three $p$-electrons such a state is constructed in the only way: the angular part of the coordinate wave function should be proportional to the mixed product of the unit radius-vectors of the three electrons, $\mathbf{n}_1[\mathbf{n}_2 \times \mathbf{n}_2]$, i.e., it should be a scalar. In other words, this state is the only one: $^4S_{3/2}$.

Then, the state with $L = 3$ in the discussed configuration should be totally symmetric in the space variables, which is immediately clear again for its components with $L = \pm 3$, where the projections of all three orbital angular momenta should be the same. But then the spin function of this state should be, correspondingly, totally antisymmetric, which is, obviously, impossible with only two projections for the spin of each electron. Therefore, in the configuration $p^3$ the $F$-state is absent.

In line with the already mentioned $^4S_{3/2}$ state, in the configuration $p^3$ there are the states $^2D_{3/2,\,5/2}$ and $^2P_{1/2,\,3/2}$, where both the coordinate function and the spin one are neither totally symmetric nor totally antisymmetric. By virtue of Hund's rules, all possible states ordered in accordance with the energy increase are

$$^4S_{3/2}; \quad ^2D_{3/2,\,5/2}; \quad ^2P_{1/2,\,3/2}\,. \tag{3.42}$$

We note that since the $p$-shell is filled here by half, the spin–orbit interaction to first order does not lift the degeneracy in the total angular momentum $J$ in the $^2D$- and $^2P$-states.

Let us consider now the opposite limiting case of $jj$-coupling for the configuration $p^3$. Obviously, the state $(1/2\ 1/2\ 1/2)$ is forbidden by the Pauli principle. So, we consider the next state $(1/2\ 1/2\ 3/2)$. Since, by virtue of the same principle, two angular momenta $1/2$ add up to $0$ only, the discussed state can only have the total angular momentum $J = 3/2$.

It was demonstrated in the previous subsection that two angular momenta $3/2$ can add up only to $0$ and $2$. Then, it can be easily seen that for the state $(1/2\ 3/2\ 3/2)$ the total angular momentum can be equal to $1/2$, $3/2$, and $5/2$.

And at last, the state $(3/2\ 3/2\ 3/2)$. Again by virtue of the Pauli principle, if the projection of the total angular momentum of one electron is $+3/2$, then for two other projections the single combination $+1/2, -1/2$ is possible. Therefore, the maximum projection of the total angular momentum is here $J_z = +3/2$. Correspondingly, for the total angular momentum the maximum value is the same: $J = 3/2$. Just the same, the projection $J_z = +1/2$ of the total angular momentum arises here in only one combination: $+3/2, +1/2, -3/2$. Since one such state should be already ascribed to $J = 3/2$, the state $(3/2\ 3/2\ 3/2)_{3/2}$ is here the single one.

Thus, in the limit of the $jj$-coupling, the configuration $p^3$ possesses the following states:

$$(1/2\ 1/2\ 3/2)_{3/2};\ (1/2\ 3/2\ 3/2)_{1/2},\ (1/2\ 3/2\ 3/2)_{3/2},\ (1/2\ 3/2\ 3/2)_{5/2};$$

$$(3/2\ 3/2\ 3/2)_{3/2}. \tag{3.43}$$

Their energies increase together with the number of electrons with $j = 3/2$; the three states $(1/2\ 3/2\ 3/2)_{1/2,\,3/2,\,5/2}$ stay degenerate in the limit of the $jj$-coupling.

Certainly, the number of the states with given total angular momentum $J$ here is also the same in the $LS$- and $jj$-schemes of addition for angular momenta. Then, the states with total angular momenta $1/2$ and $5/2$, each of which occurs only once in both schemes, coincide:

$$^2D_{5/2} = (1/2\ 3/2\ 3/2)_{5/2}; \quad ^2P_{1/2} = (1/2\ 3/2\ 3/2)_{1/2}.$$

And at last, from the comparison of (3.42) and (3.43) it follows that in the case intermediate between the two limiting ones, $LS$ and $jj$, both the ground state and the highest one have the angular momentum $J = 3/2$, and the state with $J = 1/2$ is the one closest to the highest level. As to the relative position of the last two levels with angular momenta $3/2$ and $5/2$, it cannot be fixed with such qualitative arguments. The real succession of the levels belonging to the ground state configuration $p^3$ in bismuth is as follows:

$$J = 3/2; \quad J = 3/2; \quad J = 5/2; \quad J = 1/2; \quad J = 3/2.$$

Here the states with $J = 3/2$ are orthogonal linear combinations of $^4S_{3/2}$, $^2D_{3/2}$, and $^2P_{3/2}$ in the $LS$-scheme, or $(1/2\ 1/2\ 3/2)_{3/2}, (1/2\ 3/2\ 3/2)_{3/2}$, and $(3/2\ 3/2\ 3/2)_{3/2}$ in the $jj$-scheme. Let us note that the separation between the middle level with $J = 3/2$ and the next one with $J = 5/2$ (we recall that their relative position could not be determined with qualitative arguments) proves to be the least one in the considered configuration. In other words, these two levels are closer indeed to degeneracy than all the rest.

## 3.6 Hydrogen Atom in Strong Magnetic Field

As distinct from other problems considered in the present chapter, this one requires real calculations. This problem was solved for the first time by *R.J. Elliott* and *R. Loudon* (1960). The solution presented below [9] looks like the most direct, transparent, and complete one. Though it requires some calculations, one may hope that these properties justify its presentation in this chapter which is on the whole of more qualitative character.

We start with an obvious observation: in a sufficiently strong magnetic field $B$ (the exact criteria are discussed below) the motion of an atomic electron becomes almost one-dimensional, along the magnetic field, and is described by the Coulomb potential $-e^2/z$.

The corresponding one-dimensional wave equation is

---

[9]It was given by *I.B. Khriplovich* and *G.Yu. Ruban* (2004).

$$u'' + \left( -\frac{1}{4} + \frac{n^*}{z} \right) u = 0. \tag{3.44}$$

We have introduced in it the usual dimensionless variable:

$$\frac{2z}{an^*} \longrightarrow z \,,$$

here $a = \hbar^2/m_e e^2$ is the Bohr radius, $m_e$ is the electron mass, $n^*$ is the effective quantum number, related to the electron energy as

$$E_{n^*} = -\frac{m_e e^4}{2\hbar^2 n^{*2}} \,. \tag{3.45}$$

Equation (3.44) coincides exactly with the radial equation for the $s$-wave in the three-dimensional Coulomb potential $-e^2/r$, and therefore has the common hydrogen spectrum

$$n^* = n = 1, 2, 3, \,... \,, \quad E_n^- = -\frac{m_e e^4}{2\hbar^2 n^2} \,, \tag{3.46}$$

and the set of solutions

$$u_n^-(z) = \exp\left(-z/2\right) z \, \Phi(-n, 2; z) \,, \tag{3.47}$$

where $\Phi$ is the confluent hypergeometric function (here and below we are not interested in the normalization factors). These solutions vanish at the origin. They are continued in the natural way to $z < 0$, so that the resulting solutions on the whole $z$ axis are odd under the transformation $z \to -z$ (as reflected by the superscript "minus" in (3.46), (3.47)).

There is, however, an essential difference between the present problem and the $s$-wave Coulomb one. In the last case (3.47) is the only solution. The reason is in fact as follows. One may expect in a naïve way that the radial wave equation for $R(r)(= u(r)/r)$ has two independent solutions, which behave for $r \to 0$ as $R \sim \mathrm{const}$ ($u \sim r$) and $R \sim 1/r$ ($u \sim \mathrm{const}$), respectively; both these functions are normalizable. In fact, however, $R \sim 1/r$ is no solution at all for the homogeneous wave equation if the point $r = 0$ is included in consideration, since $\Delta(1/r) = -4\pi\delta(\mathbf{r})$. This is why usually the second solution is rejected. As to our problem, equation (3.44) does not describe really the vicinity of $z = 0$, since therein we have to consider seriously the magnetic field itself. Therefore, there are no reasons here to discard those solutions of (3.44) which tend to a constant for small $z$ (and of course decrease exponentially for $z \to \infty$).

Such solutions for our problem at $z > 0$ are conveniently presented as follows [10]:

---

[10] See: I.S. Gradshteyn and I.M. Ryzhik, *Table of integrals, series and products.* New York, Academic Press, 1994. We use formula 9.327.1 from this book for $\mu = 1/2$, omit in this formula the overall factor $1/\Gamma(1/2 + \mu - \lambda)$ inessential for our purpose (the notations here are the same as in the quoted book) and rewrite the series at $\ln z$ in a compact form, as a confluent hypergeometric function.

$$u_{n^*}^+(z) = \exp\left(-z/2\right)\left\{1 - n^*z\left(\ln z\, F(1-n^*,2;z)\right.\right.$$

$$\left.\left.+ \sum_{k=0}^{\infty} \frac{\Gamma(1-n^*+k)\left[\psi(1-n^*+k) - \psi(k+2) - \psi(k+1)\right]}{\Gamma(-n^*)\,k!\,(k+1)!}\,z^{k+1}\right)\right\}; \quad (3.48)$$

here $\psi(\alpha)$ is the logarithmic derivative of the gamma function: $\psi(\alpha) = \Gamma'(\alpha)/\Gamma(\alpha)$. Functions (3.48) are in a natural way continued to $z < 0$, so that the resulting solutions on the whole $z$ axis are even under the transformation $z \to -z$ (as reflected by the superscript "plus" in the eigenfunctions (3.48) and in the corresponding eigenvalues obtained below).

Under any reasonable regularization of the logarithmic singularity at $z = 0$, first derivative of the even solutions should vanish at $z \to 0$. From this condition we obtain the following equation for the eigenvalues of $n^*$:

$$\ln\frac{a}{a_B} = \frac{1}{2n^*} + \psi(1-n^*). \qquad (3.49)$$

Here the formal logarithmic divergence at $z \to 0$ is cut off at the typical radius $a_B$ of electron orbits in the magnetic field $B$, $a_B \sim \sqrt{\hbar c/eB}$ [11]. We solve equation (3.49) in the logarithmic approximation, i.e., under the assumption

$$\lambda = \ln a/a_B \gg 1. \qquad (3.50)$$

This assumption has allowed us also, in the derivation of equation (3.49), to omit in its rhs the term $1 - \psi(1) - \psi(2) = -2\psi(1))$, which is on the order of unity.

For small $n^*$, the rhs of equation (3.49) is dominated by $1/(2n^*)$, so that the smallest root of this equation is

$$\nu_0^+ = \frac{1}{2\lambda}, \qquad (3.51)$$

and the ground state energy equals

$$E_0^+ = -\frac{m_e e^4}{2\hbar^2}\,\ln^2\left(\frac{\hbar^3 B}{m_e^2 e^3 c}\right). \qquad (3.52)$$

If $n^*$ is not small, then terms $1/(2n^*)$ in the rhs of equation (3.49) can also be neglected. In this case the eigenvalues of $n^*$ are close to the poles of the function $\psi(1-n^*)$. It is known [12] that the poles of the function $\psi(x)$ are situated at negative integer $n$, starting from zero, and the residue in each of

---

[11]See for instance: L.D. Landau and E.M. Lifshitz, *Quantum Mechanics*. §112, Problem 1.

[12]See: I.S. Gradshteyn and I.M. Ryzhik, *Table of integrals, series and products*. New York, Academic Press, 1994.

them is $-1$. In other words, for $n^* \simeq n + \varepsilon$, $n = 1, 2, 3, ...$, the function $\psi$ behaves as follows:

$$\psi(1 - n^*) \simeq \frac{1}{\varepsilon}.$$

Thus, in the vicinity of a positive integer $n^*$, equation (3.49) reduces to

$$\lambda = \frac{1}{\varepsilon}.$$

As a result, the roots of equation (3.49) discussed here are

$$n_n^{*+} = n + \frac{1}{\lambda}, \quad n = 1, 2, 3, ... , \tag{3.53}$$

and the corresponding energies equal

$$E_n^+ = -\frac{m_e e^4}{2\hbar^2 n^2} \left[ 1 - \frac{4}{n} \ln^{-1} \left( \frac{\hbar^3 B}{m_e^2 e^3 c} \right) \right]. \tag{3.54}$$

To summarize, the spectrum of the hydrogen atom in a strong magnetic field looks as follows. Each Landau level in this field serves as an upper limit of a succession of discrete levels of the Coulomb problem for the motion along the $z$ axis. This discrete spectrum consists of a singlet ground state, with the energy given by formula (3.52), and of close doublets of odd and even states with energies given by formulas (3.46) and (3.54). Over each Landau level, there is also a continuous spectrum corresponding to the motion along the $z$ axis.

This picture is valid for sufficiently low Landau levels, as long as the radius of a magnetic orbit is much less than the Bohr radius. Obviously, in a strong magnetic field this description fails for large magnetic quantum numbers, i.e., in the semiclassical region. Here the orbit radius can estimated from the well-known solution of the problem of an electron in a magnetic field [13]. The electron spectrum looks as follows:

$$E = \hbar \frac{eB}{m_e c} (N + 1/2), \quad N = n_\rho + \frac{m + |m|}{2}, \tag{3.55}$$

where $n_\rho$ is the radial quantum number for the motion in the $xy$ plane, and $m$ is the angular momentum projection onto the $z$ axis. The semiclassical estimate for the magnetic radius is

$$a_B(N) \approx \sqrt{\frac{\hbar c}{eB} N} = a_B \sqrt{N}.$$

Thus this picture of levels is valid quantitatively as long as

---

[13] See for instance: L.D. Landau and E.M. Lifshitz, *Quantum Mechanics*. §112, Problem 1.

$$\lambda = \ln \frac{a}{a_H} \gg \ln N. \tag{3.56}$$

At last, let us consider the correspondence between the obtained system of levels in a strong magnetic field and the hydrogen spectrum in the limit of vanishing external field. An elegant solution of this problem belongs to *W.H. Kleiner* (1960).

The crucial observation here is as follows. While changing the magnetic field from a vanishingly small to a very strong one, the number of nodal surfaces of a given wave function remains the same. A hydrogen wave function (in zero magnetic field) with quantum numbers $n$, $l$, $m$ has $n_r = n - l - 1$ nodal spheres, where the radial wave function $R_{nl}(r)$ turns to zero, and $l - |m|$ nodal cones with their axes directed along $z$, where the angular wave function $Y_{lm}(\theta, \phi)$ (to be more precise, the associated Legendre polynomial $P_l^{|m|}(\cos\theta)$) turns to zero.

With the increase of the magnetic field, the nodal spheres become ellipsoids of rotation, more and more prolate, tending to cylinders in the limit of infinite field. The correspondence between $n_r$ and the quantum number $n_\rho$, which refers to the motion in the $xy$ plane in a strong magnetic field, becomes obvious from this picture:

$$n_\rho = n_r. \tag{3.57}$$

Let us go over to the evolution of the hydrogen nodal cones. Due to both equation (3.57) and the conservation of the total number of nodal surfaces, the number $n_z$ of the nodes of a solution of equation (3.44) should coincide with the number of nodes of the corresponding associated Legendre polynomial $P_l^{|m|}(\cos\theta)$),

$$n_z = l - |m|. \tag{3.58}$$

In other words, $l - |m|$ nodal cones of a hydrogen wave function evolve into $n_z$ planes of constant $z$ corresponding to the nodes of an eigenfunction of equation (3.44).

Of course, the projection $m$ of the angular momentum onto the direction of magnetic field retains its meaning in the course of the whole evolution of this field.

Let us consider, for instance, the ground state in the magnetic field, with $N = 0$ (see (3.55)). Obviously, it is degenerate, and its corresponding magnetic wave functions have $n_\rho = 0$ and $m = 0, -1, -2 \dots$. We confine further to its lowest sublevel, corresponding to the ground state of equation (3.44), with $n_z = 0$. By virtue of the above arguments, the hydrogen ancestors of these wave functions should have $n_r = 0$ and $l = |m|$. In other words, these ancestors are:

$$1s; \quad 2p, \, m = -1; \quad 3d, \, m = -2; \quad \text{and so on.}$$

In fact, the ratio $\hbar^3 B/(m_e^2 e^3 c)$ becomes comparable to unity only for very strong magnetic fields, of about $10^9$ gauss. Of course, even much higher magnetic fields are required for the logarithm of this ratio to be much larger than unity. However, in semiconductors with a small effective mass of current carriers, for the electron–hole bound states the corresponding critical magnetic field turns out to be much lower. For example, in InSb it constitutes 1900 gauss only.

# 4

# Deuteron—The Hydrogen Atom of Nuclear Physics

It is known that Fermi usually started the investigation of a difficult problem with the question: "What plays the part of a hydrogen atom for this problem?" As for nuclear physics, the answer to this question causes no doubts: its hydrogen atom is the deuteron. It is surprising how many nontrivial problems related to the deuteron can be solved by means of sufficiently simple, sometimes truly back-of-the-envelope analytical calculations. Some of them are considered in this chapter.

## 4.1 Neutron–Proton System in Zero-Range Approximation

The deuteron is the simplest of nuclei, except neutron and proton, $n$ and $p$. This is the bound $np$ state with the orbital angular momentum $L = 0$ and total spin $S = 1$. In fact, the deuteron wave function contains, in line with $^3S_1$, a small admixture of $^3D_1$, which will be neglected in our discussion. The deuteron binding energy, $\varepsilon = 2.23$ MeV, is anomalously small on the nuclear scale. As a result, its wave function decreases very slowly beyond the range of nuclear forces.

Indeed, beyond this range the deuteron wave function satisfies the free wave equation

$$-\frac{1}{2\mu}\frac{1}{r}\frac{d^2(r\psi)}{dr^2} = -\varepsilon\psi. \tag{4.1}$$

Here $\mu = m_p/2$ is the reduced mass of the $np$ system; the masses of the neutron and proton are put coinciding, $m_n = m_p$, and the Planck constant $\hbar$ is put equal to unity. The normalizable solution of this equation is

$$\psi \sim \frac{e^{-\varkappa r}}{r}; \quad \varkappa = \sqrt{m_p\varepsilon} = 45.7 \text{ MeV}. \tag{4.2}$$

The characteristic distance $\varkappa^{-1} = 4.3$ fm [1], at which solution (4.2) falls down, is much larger than the range of nuclear forces $r_0 \simeq 1.2$ fm. This allows one to use for the deuteron wave function the so-called zero-range (of nuclear forces) approximation where this function is approximated by its asymptotic expression (4.2) and equals

$$\psi_d = \sqrt{\frac{\varkappa}{2\pi}} \frac{e^{-\varkappa r}}{r} ; \qquad (4.3)$$

the factor $\sqrt{\varkappa/2\pi}$ in this expression guarantees the correct normalization: $\int d\mathbf{r}\, \psi_d^2 = 1$.

The wave function of the continuous spectrum of $np$ system beyond the range of nuclear forces is written as usual in the form

$$\psi(\mathbf{k}, \mathbf{r}) = e^{ikz} + f(k, \theta) \frac{e^{ikr}}{r} ,$$

where $\theta$ is the scattering angle, and $f(k,\theta)$ is the scattering amplitude. In the low-energy limit (i.e., at small $k$), which we are interested in, scattering occurs in the $s$-state mainly (so that the dependence on $\theta$ disappears), and the wave function of continuous spectrum simplifies to

$$\psi = 1 - \frac{\alpha}{r} , \qquad (4.4)$$

where $\alpha = - \lim f(k)|_{k \to 0}$ is the so-called scattering length.

Correspondingly, the wave function of the triplet $^3S_1$ state of continuous spectrum of the neutron–proton system in the low-energy limit is

$$\psi_{St} = 1 - \frac{\alpha_t}{r} , \qquad (4.5)$$

where $\alpha_t = 5.42$ fm is the triplet scattering length. The analogous expression for the singlet $^1S_0$ function of continuous spectrum is

$$\psi_{Ss} = 1 - \frac{\alpha_s}{r} . \qquad (4.6)$$

The singlet scattering length is negative and very large: $\alpha_s = -23.7$ fm. The subscript $S$ at the wave function here and below denotes the $S$-wave, the subscripts $t$ and $s$ denote triplet and singlet states, correspondingly. We note that not only $\alpha_s$, but $\alpha_t$ as well, are much larger then the typical range of nuclear forces $r_0 \simeq 1.2$ fm.

In the zero-range approximation, the deuteron binding energy is directly related to the triplet scattering length. Indeed, from the orthogonality of triplet wave functions (4.3) and (4.5), which refer to different energies, the relation

---

[1] 1 fm (fermi) $= 10^{-13}$ cm.

$$\alpha_t = \frac{1}{\varkappa} \qquad (4.7)$$

follows immediately. It is valid with accuracy $\sim 20\%$ (compare the numerical values $\alpha_t = 5.4$ fm and $\varkappa^{-1} = (m_p\varepsilon)^{-1/2} = 4.3$ fm).

Let us note that in a system with a negative scattering length in general, and in the singlet $np$ state in particular, bound states cannot exist. Indeed, the lowest of the bound states has no nodes, i.e., its radial wave function is of definite sign. On the other hand, this radial function should be orthogonal to function (4.4), which is certainly impossible with $\alpha < 0$ when the last function is of definite sign also.

If $\alpha > 0$, then formally the wave function of a bound state in the form $\exp(-r/\alpha)/r$, orthogonal to (4.4), always exists. In fact, however, this function can serve as a true solution of the problem only under condition $\alpha \gg r_0$. Therefore, to guarantee the existence of a bound state, the scattering length should not only be positive, but also much larger than the range of potential. Otherwise, a bound state may or may not exist.

## 4.2 Radiative Capture of Neutron by Proton

We consider now the reaction $np \to d\gamma$. As usual, this electromagnetic transition is dominated by the lowest multipolarities, $E1$ and $M1$. Let us compare their intensities for slow neutrons, such that their energy is much smaller than $\varepsilon$, and correspondingly, the relative momentum $p$ in the initial state is small as compared to $\varkappa$.

The operator of $E1$ transition $V_E = -e\,\mathbf{r}\,\mathbf{E}$ does not change the total spin of the system, so that this transition proceeds from the initial $^3P$ state. In the zero-range approximation, the wave function of this state can be considered as a free one. The corresponding matrix element is

$$< {}^3P|\mathbf{r}|{}^3S > = \int d\mathbf{r}\, e^{-i\mathbf{p}\mathbf{r}}\mathbf{r}\sqrt{\frac{\varkappa}{2\pi}}\frac{e^{-\varkappa r}}{r} = -i\,4\sqrt{2\pi}\,\alpha_t^{7/2}\,\mathbf{p} \qquad (4.8)$$

(for arbitrary $\mathbf{p}$ and spherically symmetric function $^3S$, vector $\mathbf{r}$ "chooses" by itself the $P$-wave from $e^{-i\mathbf{p}\mathbf{r}}$; to our approximation, it is sufficient to put at once $e^{-i\mathbf{p}\mathbf{r}} = 1 - i\mathbf{p}\mathbf{r}$).

Let us go over now to the $M1$ operator. One may omit in it the orbital momentum contribution, since we neglect the $D$-wave admixture. Thus, the $M1$ operator reduces to

$$V_M = -\frac{e}{2m_p}\left(\mu_p\boldsymbol{\sigma}_p + \mu_n\boldsymbol{\sigma}_n\right)\mathbf{B}. \qquad (4.9)$$

Here $\boldsymbol{\sigma}_p$ and $\boldsymbol{\sigma}_n$ are the spin operators of the proton and neutron, respectively; $\mu_p = 2.79$ and $\mu_n = -1.91$ are their magnetic moments; the velocity of light $c$ is put equal to unity, as well the Planck constant $\hbar$. The spin operator in this expression is conveniently rewritten as follows:

$$\mu_p\boldsymbol{\sigma}_p + \mu_n\boldsymbol{\sigma}_n = \frac{1}{2}(\mu_p + \mu_n)(\boldsymbol{\sigma}_p + \boldsymbol{\sigma}_n) + \frac{1}{2}(\mu_p - \mu_n)(\boldsymbol{\sigma}_p - \boldsymbol{\sigma}_n). \quad (4.10)$$

The structure $\frac{1}{2}(\boldsymbol{\sigma}_p + \boldsymbol{\sigma}_n)$ in this expression is the total spin operator of the nucleons, which might couple the deuteron only with the $^3S_1$ state of the continuous spectrum. However, the coordinate wave functions of the deuteron and this initial state are orthogonal, so that the total spin operator is not operative in the present case. Thus, the $M1$ transition from the singlet $^1S_0$ state of the continuous spectrum proceeds due to the operator

$$\tilde{V}_M = -\frac{e}{4m_p}(\mu_p - \mu_n)(\boldsymbol{\sigma}_p - \boldsymbol{\sigma}_n)\mathbf{B}. \quad (4.11)$$

The radial matrix element reduces here to the overlap integral of the wave functions (4.3) and (4.6):

$$\sqrt{\frac{\varkappa}{2\pi}} \int_0^\infty 4\pi r^2 dr \left(1 - \frac{\alpha_s}{r}\right)\frac{e^{-r/\alpha_t}}{r} = \sqrt{8\pi\alpha_t}(\alpha_t - \alpha_s). \quad (4.12)$$

Let us compare now the intensities of the $E1$ and $M1$ transitions. Since the electric and magnetic fields of emitted photon are equal by modulus, we have to estimate only the ratio of matrix elements (4.8) and (4.12) (the last one should be additionally multiplied by $(\mu_p - \mu_n)/(4m_p)$). An elementary estimate demonstrates that the $M1$ transition dominates as long as the relative momentum in the center-of-mass frame is bounded by condition

$$p^2 \lesssim \frac{(\mu_p - \mu_n)^2(\alpha_t - \alpha_s)^2}{64\alpha_t^6 m_p^2}.$$

The corresponding bound on the neutron energy (in the laboratory frame) is

$$E_n \lesssim 0.1 \text{ MeV}. \quad (4.13)$$

Just this region of slow neutrons, where the $M1$ transition dominates, is considered below.

When calculating the cross-section of radiative capture, one can do without the rather tedious standard technique of Clebsch–Gordan coefficients. To this end, let us consider the contribution of operator (4.11) directly to the matrix element squared, summed over the projections $m$ of the deuteron spin. Using the completeness relation for the deuteron wave functions, we obtain

$$\sum_m \langle {}^1S_0| (\boldsymbol{\sigma}_p - \boldsymbol{\sigma}_n)\mathbf{B}^*|{}^3S_1, m\rangle\langle {}^3S_1, m|(\boldsymbol{\sigma}_p - \boldsymbol{\sigma}_n)\mathbf{B}|{}^1S_0\rangle$$

$$= \langle {}^1S_0| [(\boldsymbol{\sigma}_p - \boldsymbol{\sigma}_n)\mathbf{B}]^2|{}^1S_0\rangle.$$

By virtue of spherical symmetry of the $|{}^1S_0\rangle$ state, this expression reduces to

$$\frac{1}{3}|\mathbf{B}|^2\langle {}^1S_0|(\boldsymbol{\sigma_p} - \boldsymbol{\sigma_n})^2|{}^1S_0\rangle = \frac{1}{3}|\mathbf{B}|^2\langle {}^1S_0|6 - 2\boldsymbol{\sigma_p\sigma_n}|{}^1S_0\rangle.$$

We note further that $|^1S_0\rangle$ is eigenstate of operator

$$(\sigma_\mathbf{p} + \sigma_\mathbf{n})^2 = 6 + 2\sigma_\mathbf{p}\sigma_\mathbf{n}$$

with eigenvalue 0. And then

$$(6 - 2\sigma_\mathbf{p}\sigma_\mathbf{n})|^1S_0\rangle = 12|^1S_0\rangle\,.$$

Thus, the discussed matrix element squared is equal to $4|\mathbf{B}|^2$. Since the vector potential of emitted photon is

$$\mathbf{A}^* = \sqrt{\frac{4\pi}{2\omega}}\,\mathbf{e}^*\,e^{i\omega(t - i\mathbf{nr})}\,, \qquad \mathbf{n} = \mathbf{k}/k\,, \quad (\mathbf{ne}) = 0\,,$$

we have $4|\mathbf{B}|^2 = 8\pi\omega$. Now we sum over two possible polarizations $\lambda$ of the emitted photon, and obtain

$$\sum_{m,\lambda} \langle^1S_0|\,(\sigma_p - \sigma_n)\,\mathbf{B}^\lambda|^3S_1, m\rangle\langle^3S_1, m|\,(\sigma_p - \sigma_n)\,\mathbf{B}^\lambda|^1S_0\rangle$$

$$= 16\pi\omega = 16\pi\varepsilon\,. \tag{4.14}$$

Then, the radial matrix element (4.12) squared is

$$8\pi\alpha_t(\alpha_t - \alpha_s)^2\,.$$

At last, we have to take into account the phase volume of the final state

$$2\pi \int \delta(\varepsilon - \omega)\,\frac{4\pi\omega^2 d\omega}{(2\pi)^3} = \frac{\varepsilon^2}{\pi}\,, \tag{4.15}$$

and to divide the intermediate result by the current density of incoming neutrons $v$. Besides, we should recall that a reaction with unpolarized neutron and proton is considered, so that all four possible $np$ states with the orbital angular momentum 0: $^1S_0$ and $^3S_1$, $m = -1, 0, 1$, are of the same probability [2]. Since only one of them serves as an initial one, the result should be divided by 4.

Finally, after simple calculations we arrive at the following expression for the total cross-section of the neutron radiative capture:

$$\sigma_{rc} = 2\pi\,\alpha\,(\mu_p - \mu_n)^2 \left(1 - \frac{\alpha_s}{\alpha_t}\right)^2 \left(\frac{\varepsilon}{m_p}\right)^{3/2} \frac{1}{m_p^2}\frac{1}{v}\,, \tag{4.16}$$

---

[2] Of course, one can obtain the same number of possible spin states for two unpolarized particles with spin 1/2 otherwise, independently of their orbital angular momentum. It is sufficient to count all possible combinations of their spin projections: $(+\,+), (+\,-), (-\,+), (-\,-)$.

here $\alpha = e^2$. Let us note that for the so-called thermal neutrons, with energy $\lesssim 0.1$ eV, this cross-section is huge, due to the factor $1/v$ ($c/v$ in the common units).

Exactly in the same way one can find the total cross-section of the deuteron photodissociation which is the reaction inverse with respect to the considered radiative neutron capture. The cross-sections of the two reactions are related through the principle of detailed balance. In other words, they differ, first, by the number of spin final states: 3 possible deuteron polarizations and 2 photon polarizations, altogether 6 final states in $np \rightarrow d\gamma$; 2 possible polarizations both for the proton and neutron, altogether 4 final states in $\gamma d \rightarrow np$. Then, the phase volumes of the final states: $\varepsilon^2/\pi$ in $np \rightarrow d\gamma$ (see (4.15)) and

$$2\pi \int \delta \left( \omega - \varepsilon - 2 \cdot \frac{p^2}{2m_p} \right) \frac{d\mathbf{p}}{(2\pi)^3} = \frac{m\sqrt{m(\omega - \varepsilon)}}{2\pi} \qquad (4.17)$$

in $\gamma d \rightarrow np$. And finally, the initial fluxes: $v$ in $np \rightarrow d\gamma$ and 1 ($c$ in common units) in $\gamma d \rightarrow np$. With the account for all these factors, the total cross-section of the deuteron photodissociation equals (E. Fermi, 1935)

$$\sigma_{phd} = \frac{2\pi}{3} \alpha \left(\mu_p - \mu_n\right)^2 \left( 1 - \frac{\alpha_s}{\alpha_t} \right)^2 \sqrt{\frac{\omega - \varepsilon}{\varepsilon}} \frac{1}{m_p^2}. \qquad (4.18)$$

In conclusion, let us discuss briefly the radiative capture of polarized thermal neutrons. Will the produced photon be polarized in this case? Within the approximation applied here, it will not. Since the initial state $^1S_0$ is totally spherically symmetric, the polarization of an initial particle is not transferred to a final one. However, due to the admixture of $D$ state in the deuteron wave function and in the incoming wave, the $M1$ transition proceeds also from the triplet initial state. Besides, due to the $D$ wave admixture, the $E2$ transition becomes possible as well. And at last, relativistic corrections to the $M1$ operator, responsible for nonadditivity of the nucleon magnetic moments in deuteron, also make possible the magnetic dipole transition from the triplet initial state. All these effects are tiny, but their investigation gives information on delicate details of the $np$ interaction at low energies.

## 4.3 Three-Body Problem in Zero-Range Approximation

We address now the three-body problem, under the same assumption that the scattering lengths for any pair of the bodies are large as compared to the range of potential. If the two-body scattering lengths are positive, so that any two of the particles can produce a bound state, then it is only natural that there are bound states in the three-particle system as well. The remarkable observation is that even for negative scattering lengths as well (when there are no two-body bound states!) in a three-body system bound states may arise (V. N. Efimov, 1970).

For the beginning, we present a very qualitative, but instructive argument in favor of this assertion. Let the scattering length of particle 1 on each of particles 2 and 3 be large (though negative). It means that particle 1 resides for a long time in shallow potential wells near each of particles 2 and 3 (though does not form a bound state with either of these particles separately). Then, an additional attraction arises between particles 2 and 3, as a result of tunneling of particle 1 between these two wells [3]. Of course, generally speaking, particle 1 is in no way singled out as compared to particles 2 and 3, so that a similar attraction arises between 3 and 1, and between 1 and 2. This is a collective three-particle effect.

Now we are going over to more quantitative estimates. We assume for simplicity sake that the particle masses are the same, $m_1 = m_2 = m_3 = m$, and the two-body scattering lengths are the same as well, $\alpha_{12} = \alpha_{23} = \alpha_{31} = \alpha$. Let us estimate with qualitative arguments the effective potential $U$ of the discussed system. In our zero-range approximation this potential should be independent of $r_0$. On the other hand, we assume that the (negative) scattering length is large, $|\alpha| \gg 1$, so that it also should not enter the expression for $U$. Then the dimensional arguments result in the following expression:

$$U = -\xi \frac{\hbar^2}{mR^2} . \tag{4.19}$$

It is natural to consider the discussed potential as an attractive one, so that the dimensionless constant $\xi$ is positive. As to an effective length $R$ in this expression, it should be a symmetrical function of all three coordinates. Besides, since expression (4.19) describes an essentially three-body effect, it should vanish when one of the particles goes to infinity. Then the simplest expression for $R^2$ is

$$R^2 = r_{12}^2 + r_{23}^2 + r_{31}^2 .$$

At the vanishing energy and $\xi > 1/4$, the wave function, which describes the motion along the generalized coordinate $R$, can be written as [4]

$$\Psi \sim \frac{1}{\sqrt{R}} \cos \left( \sqrt{\xi - 1/4} \, \ln \frac{R}{r_0} + \text{const} \right) . \tag{4.20}$$

This expression refers to the region where all relative coordinates $r_{ij}$, and hence $R$ as well, exceed the range of $r_0$ of two-body potentials. On the other hand, the effective potential (4.19) was constructed under the assumption that the scattering lengths are large, i.e., that $R < |\alpha|$. Therefore, the number of nodes $n$ of wave function (4.20) in the interval $r_0 < R < a$, i.e., the number of levels, of bound states with negative energy, is

---

[3]Let us recall in this connection the molecular hydrogen ion $H_2^+$. Here the bound state forms due to the tunneling of the electron (particle 1) between the potential wells near two nuclei (particles 2 and 3). However, this example differs from our problem essentially since each nucleus by itself bounds the electron.

[4]See: L. D. Landau and E. M. Lifshitz, *Quantum Mechanics.* §35.

$$n = \frac{1}{\pi} \sqrt{\xi - 1/4} \, \ln \frac{a}{r_0} \, . \tag{4.21}$$

Whether condition $n > 1$, under which these levels really exist, is valid, depends on the values of ratio $a/r_0$ and of parameter $\xi$, i.e., on the concrete dynamics of the problem. At any rate, this is a problem of a very delicate effect, of states that are bound very weakly. Therefore, the experimental observation of such states and the proof that their nature is due indeed to the discussed effect, is extremely difficult. No wonder therefore that the first convincing experimental proof of the existence of the Efimov effect was obtained only recently, though not in a system of nucleons, but in the gas of ultracold cesium atoms (*T. Kraemer* et al., 2006).

In conclusion, let us come back to potential (4.19) and wave function (4.20). If this potential is applicable for arbitrary small $R$, i.e., if the range $r_0$ of two-body potentials tends to zero, then solution (4.20) describes in fact in this limit fall to the center [5]. In other words, the three-body problem is unstable in this limit, it has no ground state (*L. Thomas*, 1935).

Certainly, this assertion is directly related to nuclear physics. It is sufficient to recall, for instance, the tritium nucleus which is the bound state of a proton and two neutrons. The zero-range approximation is absolutely inapplicable for its description. Moreover, the above arguments are generalized in an elementary way for larger number of particles, $N > 3$. Thus, the deuteron is the only nucleus for which the zero-range approximation results in a stable state.

---

[5]See again: L.D. Landau and E.M. Lifshitz, *Quantum Mechanics*. §35.

# 5

---

# Semiclassical Approximation in Complex Plane

## 5.1 General Notions

As is known, the semiclassical approximation applies to the solution of the Schrödinger equation

$$-\frac{\hbar^2}{2m}\frac{d^2\psi(x)}{dx^2} + U(x)\psi(x) = E\psi(x) \qquad (5.1)$$

if the potential $U(x)$ changes slowly at the distances on the order of the de Broglie wavelength of the particle, i.e., at those distances where the solution $\psi(x)$ itself varies essentially. Equation (5.1) can be rewritten somewhat otherwise:

$$\frac{d^2\psi(x)}{dx^2} + k^2(x)\psi(x) = 0, \quad k^2(x) = \frac{2m}{\hbar^2}[E - U(x)]. \qquad (5.2)$$

Now the condition of applicability of the semiclassical approximation is that $k(x)$ should vary slowly at the distances on the order of $1/k(x)$. Equation (5.2) describes in fact a sufficiently wide class of phenomena, in no way confined to quantum mechanics. One such problem will be considered below.

In view of the universality of semiclassical approximation, it is useful to formulate its applicability limits in general form, irrespective of quantum mechanics. Let

$$k^2(x) = k_0^2\, q(y)\,, \quad y = x/R\,;$$

here $R$ is the typical distance where $k^2(x)$ varies, and $q(y)$ is comparable with unity everywhere, but the vicinity of the so-called turning point $x_0$, where $k(x)$ vanishes. Then equation (5.2) transforms to

$$\varepsilon^2\frac{d^2\psi(y)}{dy^2} + q(y)\psi(y) = 0\,, \qquad (5.3)$$

where

$$\varepsilon = 1/(k_0 R) \ll 1 .$$

Thus, the region of applicability of the semiclassical approximation is the problems which reduce to differential equations with a small coefficient at the higher derivative. If we come back to quantum mechanics, in it this parameter,

$$\varepsilon = \hbar/(p_0 R) ,$$

is nothing but the smallness of the Planck constant $\hbar$ as compared to the typical action $p_0 R$ of the considered system.

As is known, in the semiclassical approximation the independent solutions $\psi_+$ and $\psi_-$ of equation (5.2) are as follows:

$$\psi_\pm = \frac{1}{\sqrt{k(x)}} \exp\{\pm i\, \Phi(x)\}, \quad \Phi(x) = \int^x dx\, k(x) . \qquad (5.4)$$

These solutions approximate the true ones with a powerlike accuracy in $\varepsilon$ [1].

The semiclassical approximation is obviously inapplicable near the turning point $x_0$, where $k(x_0) = 0$. Meanwhile, a rather typical formulation of problem is that the semiclassical asymptotics of a solution is known on one side of the turning point $x_0$, and one has to construct with it the correct asymptotics on the other side of this point. An effective solution of such problems is based on going out of the real axis to the complex plane of independent variable. It is applicable as well when the turning points themselves lie in the complex plane. The concrete version of this technique used below goes back to A. Zwaan (1929) [2]. It looks to be the most natural and reliable one.

## 5.2 One Turning Point. Reflection from Barrier

We start with a simple and well-known problem. Let a particle be reflected off a potential barrier (see Fig. 5.1) at a point $x_0$, so that the equation $k^2(x) = 0$ has here a simple root. In the classically inaccessible region, inside the barrier, the solution decreases to the left of $x_0$. One should find the semiclassical solution to the right of this point.

Usually, this problem is solved as follows. In the vicinity of the turning point $x_0$ the potential is approximated by a linear function. Then in this region one constructs the exact solution (this is the so-called Airy function) which decreases exponentially inside the barrier, and at last the asymptotics of this solution is found to the right of $x_0$.

We will proceed otherwise. Let us choose $x_0$ as the origin, and the scale such that $k^2(z) = z$ in a sufficiently large vicinity of this origin (here, and

---

[1] See for instance: L.D. Landau and E.M. Lifshitz, *Quantum Mechanics*. §46.

[2] About forty years ago it was a sort of "common knowledge" of a group of theorists from the Institute of Nuclear Physics in Novosibirsk; among them one should mention first of all *S.S. Moiseev, R.Z. Sagdeev, A.I. Vainshtein,* and *G.M. Zaslavsky.*

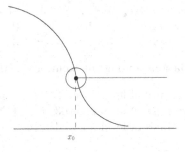

Fig. 5.1

only here, the present approach does not differ from the standard one). Then in solutions (5.4)

$$k(z) = z^{1/2}, \quad k^{-1/2}(z) = z^{-1/4}, \quad \Phi(z) = \int_0^z z^{1/2} = \frac{2}{3} z^{3/2}.$$

Thus, the semiclassical solutions (5.4) have the branching point at $z = 0$. The cut starting at it is conveniently drawn to the right along the real semi-axis $z$ (see Fig. 5.2). Obviously, both on the upper and the lower sides of the cut the phase $\Phi(z)$ of solutions $\psi_\pm$ is real. These are the so-called lines of level where both solutions are equal by modulus. Two more lines of level, with $\arg z = 2\pi/3$ and $\arg z = 4\pi/3$, start at the origin $z = 0$. Four lines of level (marked by numbers) and the cut are presented in Fig. 5.2.

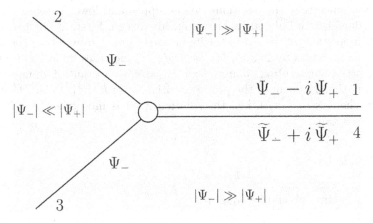

Fig. 5.2

The change of $\arg \Phi(z)$ under the transition from one line of level to another looks as follows:

| line of level | 1 | 2 | 3 | 4 |
|---|---|---|---|---|
| $\arg z$ | 0 | $2\pi/3$ | $4\pi/3$ | $2\pi$ |
| $\arg \Phi(z)$ | 0 | $\pi$ | $2\pi$ | $3\pi$ |

Hence, the following relations among the solutions hold:

between lines of level 1 and 2     $|\psi_-| \gg |\psi_+|$,
between lines of level 2 and 3     $|\psi_-| \ll |\psi_+|$,
between lines of level 3 and 4     $|\psi_-| \gg |\psi_+|$.

The solution we are interested in, is exponentially small in the classically inaccessible region, i.e., it coincides with $\psi_-$ on the negative real semi-axis. Therefore, clearly it equals to $\psi_-$ on the lines of level 2 and 3. Then, at the transition from the line 2 via the upper half-plane to the line 1 (i.e., to the upper side of the cut), the coefficient at the solution $\psi_-$, which is exponentially large on this path, remains the same. However, since the accuracy of the asymptotic solutions (5.4) themselves is only powerlike, there are no reasons to assume that the second asymptotic solution, which is exponentially small in the region between 2 and 1, will not arise under this transition [3]. We write therefore the wave function on the upper side of the cut as $\psi_- + a\,\psi_+$, where $a$ is some coefficient, unknown at the moment.

Just in the same way, at the transition from the line 3 to the line 4 (i.e., to the lower side of the cut), the wave function on this side of the cut turns out to be equal to $\tilde{\psi}_- + b\,\tilde{\psi}_+$, where $b$ is one more unknown coefficient. Tildes in the last expression have appeared due to the following reason. In fact, the origin is a regular point for our exact equation, so that the exact solution has here no cut at all, the cut arises in the semiclassical approximation only. It follows hence that the solutions on the upper and lower sides of the cut should coincide. And of course, they coincide indeed. First, at the transition from the upper side of the cut to the lower one by going around the origin, i.e., when $\arg z$ increases by $2\pi$, $\arg \Phi(z) = \arg(z^{3/2})$ increases by $3\pi$, so that the phase $\Phi(z)$ of our solution changes sign. Besides, as a result of going around the origin, the argument of the preexponential factor $k^{-1/2}(z) = z^{-1/4}$ in the solution changes by $-\pi/2$, i.e., the solution itself is multiplied by $-i$. As a result, we obtain

$$\tilde{\psi}_\pm = -i\,\psi_\mp\,.$$

Therefore, in fact,

$$\tilde{\psi}_- + b\,\tilde{\psi}_+ = -i\,\psi_+ - i\,b\,\psi_-\,.$$

And this, in turn, should be equal to

$$\psi_- + a\,\psi_+\,.$$

As a result, $a = -i$, $b = i$, and the solution on the right semi-axis is

---

[3]Change of the form of an asymptotic solution in the course of its analytic continuation from one region to another is called the Stokes phenomenon.

$$\frac{1}{\sqrt{k(z)}} \left\{ \exp\left[ -i \int_0^z k(z)dz \right] + e^{-i\pi/2} \exp\left[ i \int_0^z k(z)dz \right] \right\}.$$

On the other hand, the wave function on the left semi-axis is

$$\psi_- = \frac{1}{\sqrt{k(z)}} \exp\left[ -i \int_0^z k(z)dz \right] = \frac{e^{-i\pi/4}}{\sqrt{|k(z)|}} \exp\left[ -\int_0^z |k(z)|dz \right].$$

Thus, asymptotics on the negative left and positive right semi-axes are matched as follows:

$$\frac{e^{-i\pi/4}}{\sqrt{|k(z)|}} \exp\left[ -\int_0^z |k(z)|dz \right]$$

$$\longrightarrow \frac{1}{\sqrt{k(z)}} \left\{ \exp\left[ -i \int_0^z k(z)dz \right] + e^{-i\pi/2} \exp\left[ i \int_0^z k(z)dz \right] \right\},$$

or, in a more convenient form,

$$\frac{1}{2\sqrt{|k(z)|}} \exp\left[ -\int_0^z |k(z)|dz \right]$$

$$\longrightarrow \frac{1}{2\sqrt{k(z)}} \left\{ \exp\left[ -i \int_0^z k(z)dz + i\pi/4 \right] + \exp\left[ i \int_0^z k(z)dz - i\pi/4 \right] \right\}$$

$$= \frac{1}{\sqrt{k(z)}} \cos\left\{ \int_0^z k(z)dz - \pi/4 \right\}. \qquad (5.5)$$

In conclusion of this section, we note that in the particular case $k(z) = z^{1/2}$ the known asymptotics of the Airy function $Ai(z)$ follow from formula (5.5) [4]. For real $z$, at $z \longrightarrow -\infty$

$$Ai(z) \simeq \frac{1}{2|z|^{1/4}} \exp(-\frac{2}{3}|z|^{3/2}), \qquad (5.6)$$

and at $z \longrightarrow \infty$

$$Ai(z) \simeq \frac{1}{z^{1/4}} \cos\left( \frac{2}{3}z^{3/2} - \pi/4 \right) = \frac{1}{z^{1/4}} \sin\left( \frac{2}{3}z^{3/2} + \pi/4 \right). \qquad (5.7)$$

As distinct from the standard derivation of these asymptotics, we need here neither the integral representation of the Airy function, nor the steepest descent method. Simple algebra was in fact quite sufficient for us.

---

[4]The common form of asymptotics (5.6), (5.7) is somewhat different; it corresponds to the situation when the classically inaccessible region lies to the right of the turning point, but not to the left, as in our derivation.

## 5.3 Two Turning Points. Bohr–Sommerfeld Quantization Rule

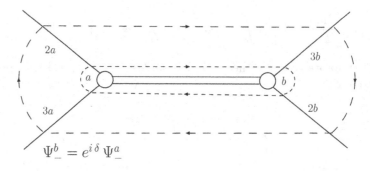

$$\Psi_-^b = e^{i\,\delta}\,\Psi_-^a$$

*Fig. 5.3*

Let us consider the problem of motion in a potential well between two turning points $a$ and $b$. In this problem a cut in the complex plane $z$ is conveniently drawn along the real axis between these points, in the classically accessible region (see Fig. 5.3). We assume that the semiclassical approximation is applicable at large distances from both turning points $a$ and $b$, but no assumption is made on its applicability in the region between these points.

On the real axis beyond the interval $ab$ the solution is exponentially small. Such semiclassical solutions to the left and to the right of this interval can be chosen as $\psi_-^a$ and $\psi_-^b$, respectively. It can be done not only on the mentioned parts of the real axis, but as well in the sectors of the complex plane including these parts (superscripts $a$ and $b$ mean that the phases $\varPhi(z)$ of the corresponding functions are counted off the points $a$ and $b$, respectively). Of course, these solutions look the same way also on the lines of level going to infinity from the points $a$ and $b$. As to other regions of the complex plane, sufficiently far from the turning points $a$ and $b$, the functions $\psi_-^a$ and $\psi_-^b$ are exponentially large. Therefore, the farther lie these regions from the turning points, the better $\psi_-^a$ and $\psi_-^b$ approximate the true solution. In fact, functions $\psi_-^a$ and $\psi_-^b$ differ by a phase factor only. We choose as an asymptotic semiclassical solution an appropriate linear combination of these functions denoted from now on by $\psi_-$ ; its concrete form is inessential for us.

Let us take now the true solution at some point $z$ and go with it around a closed contour with the cut inside it (see Fig. 5.3); for the sake of definiteness, let us go clockwise. We assume that the exact solution has no singularities inside the chosen contour. Then this solution, being single-valued, comes back to its initial value, up to an additional overall phase $2\pi n$ where $n$ is an integer.

Let this contour be at first far away from the turning points, so that the true solution on it is well approximated by the semiclassical one $\psi_-$. With this approximate solution, we squeeze now the initial contour in such a way that it goes along the upper and lower sides of the cut. Then, the phase acquired after going around the cut is

$$\delta = \oint k(z)dz = 2\int_a^b k(z)dz \, .$$

Besides, an additional phase arises due to the preexponential factor in the semiclassical solution. To take into account both turning points, we write this factor as

$$[(z-a)(z-b)]^{-1/4} \, .$$

For the chosen direction of going around the contour, the contribution of the preexponential factor to the phase is $\pi$. Thus, we arrive at the following rule:

$$\oint k(z)dz + \pi = 2\pi n \, .$$

Shifting $n$ by unity, we obtain the quantization rule in the standard form:

$$\oint k(z)dz = 2\pi\left(n + \frac{1}{2}\right), \tag{5.8}$$

or

$$\int_a^b k(z)dz = \pi\left(n + \frac{1}{2}\right) \, .$$

The accuracy of quantization rule (5.8) is determined by the accuracy with which the semiclassical function $\psi_-$ approximates the true solution. And in turn, this depends on how far away from the turning points one can go without coming across new turning points or singularities of the differential equation itself. For the oscillator potential $m\omega^2 z^2/2$, the only singular point of the differential equation is at infinity, and there are only two turning points, $a$ and $b$, just those considered when the quantization rule was derived. In this derivation, we did not assume that the semiclassical approximation was applicable in the region between the turning points. Just due to it, the quantization rule (5.8) is exact for an oscillator, it is valid not only for large $n$, but even for $n = 0$ as well.

Obviously, the existence of only two turning points is a special feature just of the oscillator potential which is quadratic in $z$. With any nonlinear addition to the oscillator potential, or for instance when changing it to $\beta z^4$, the number of turning points in the complex plane increases, so that the quantization rule (5.8) is not exact anymore.

On the other hand, while in the complex plane, at large distance from the turning points $a$ and $b$ any oscillator function is well approximated by the semiclassical solution $\psi_-$, in the region between the turning points the situation is different. Here the semiclassical asymptotics for the oscillator wave functions is valid, as usual, for large $n$ only.

## 5.4 Two Turning Points. Underbarrier Transition

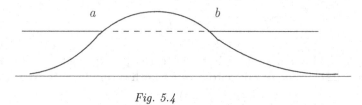

*Fig. 5.4*

Let the energy of a wave, incident from the left on a potential barrier, be less than the maximum of the latter (see Fig. 5.4). This wave is partially reflected from the barrier, but partially penetrates it and goes farther to the right out of the barrier. Correspondingly, in the complex plane $z$ there are two turning points $a$ and $b$ on the real axis (see Fig. 5.5). Let us find the coefficients

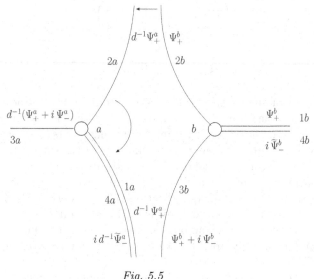

*Fig. 5.5*

of transition and reflection in the semiclassical approximation. As well as the problem of quantization considered above, this problem can be solved without the assumption that the semiclassical approximation is applicable in the region between the turning points. The only condition necessary is that this approximation applies outside, at large distances from the points $a$ and $b$. However, here this general solution turns out rather tedious. Therefore, we confine ourselves below to a detailed consideration of a more simple particular case when the semiclassical approximation applies also in the region between the turning points.

Obviously, at the exit from the barrier, to the right of the point $b$, there is a transmitted wave $\psi_+^b$ only; the coefficient at $\psi_+^b$ is made equal to unity for the time being. The cut starting at the point $b$ is chosen going to the right along the real axis, so that $\psi_+^b$ is a solution on its upper side, i.e., on the line $1b$. Since in the sector between the level lines $1b$ and $2b$ the function $\psi_+^b$ is exponentially small, this function serves as the solution on the line $2b$ as well. On the other hand, the solution on the lower side of the cut, on the line $4b$, looks as $i\psi_-^b$. The line of reasoning, completely analogous to that used in the previous section, leads to the conclusion that the solution on the line $3b$ is $\psi_+^b + i\psi_-^b$.

We note here that the solution in the complex plane, found now, is the second linearly independent solution for a single turning point (the first one was obtained in Section 5.2).

Then, the lines of level $2a$ and $2b$ coalesce at infinity (see Fig. 5.5), so that the solutions on them should coincide asymptotically. Therefore, on the line $2a$ the solution is $\psi_+^b$ as previously, though it is convenient to rewrite it, shifting the origin from $b$ to $a$:

$$\psi_+^b(z) = \exp\left[i\int_b^a k(z)dz\right]\psi_+^a(z) = d^{-1}\psi_+^a(z).$$

We recall here that, by virtue of obvious physical arguments, the coefficient

$$d = \exp\left[i\int_a^b k(z)dz\right] = \exp\left[-\int_a^b |k(z)|dz\right] \tag{5.9}$$

should be exponentially small. However, up to the preexponential factor, inessential in the present case, it coincides with $\psi_+^a(b)$. It means that in the considered sector $|\psi_+^a| \ll |\psi_-^a|$. To satisfy this inequality, we draw the cut starting at the branching point $a$, along the line of level going down from this point [5]. After it we find easily (see Fig. 5.5) the solution of interest for us on the line of level going to the left along the real axis:

$$d^{-1}\left(\psi_+^a + i\psi_-^a\right).$$

We recall here that in the sector between the lines of level $2b$ and $3b$ the solution $\psi_-^b$ is exponentially small. No wonder therefore that no trace of it is left on the lines of level $1a$ and $2a$ after the transition from the system of lines $b$ to the system of lines $a$ via the point at infinity. On the other hand, $\psi_-^a$ is exponentially large in the sector between the lines of level $2a$ and $1a$, and therefore cannot arise during this transition.

---

[5] An option is to draw the cut from the point $a$ to the left along the real axis (symmetrically with the cut drawn from the point $b$ to the right), but to choose here otherwise the sheet of the Riemann surface.

Thus, the asymptotics on the left and on the right on the real axis are matched (after additional multiplication by $d$) in the following way:

$$\psi_+^a + i\,\psi_-^a \longrightarrow d\,\psi_+^b\,.$$

This relation can be somewhat improved, if one recalls that the sum of the transition and reflection coefficients, i.e., $D = d^2$ and $R$ (which is the modulus squared of the coefficient at $\psi_-^a$), respectively, should be equal to unity. Then we find:

$$\psi_+^a + i\left(1 - \frac{1}{2}d^2\right)\psi_-^a \longrightarrow d\,\psi_+^b\,. \tag{5.10}$$

In concluding this section, we present the result for the transition and reflection coefficients, derived without the assumption of applicability of semiclassical approximation between the turning points, i.e., without the assumption that $d^2$ is exponentially small:

$$D = \frac{d^2}{1 + d^2}\,; \quad R = \frac{1}{1 + d^2}\,; \quad D + R = 1\,. \tag{5.11}$$

## 5.5 Two Turning Points. Overbarrier Reflection

Let us consider now the following problem. A wave comes (from the left) to a smooth barrier, and the energy of the wave exceeds the maximum of the barrier (see Fig. 5.6). In the classical mechanics in this case there is no reflection from the barrier at all. In fact, the whole real axis $z$ is a line of level, without any turning points on it.

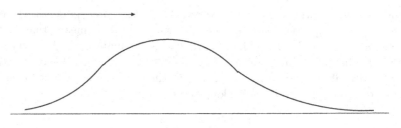

Fig. 5.6

However, in the complex plane $z$ turning points exist. Since both the potential and the particle energy are real, these turning points arise as complex conjugate pairs. We will take into account only those two turning points that are closest to the real axis (see Fig. 5.7). We assume besides that the potential tends to zero sufficiently rapidly for $\mathrm{Re}\,z \longrightarrow \pm\infty$, so that the two lines of level originating from each turning point, $a$ and $a^*$, asymptotically tend to the real axis for $\pm\infty$. The picture of lines of level presented in Fig. 5.7

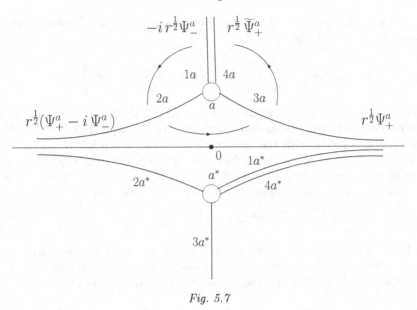

Fig. 5.7

resembles that of Fig. 5.6 which describes the underbarrier transition. The difference consists, first, in the rotation by $\pi/2$ around the origin. Then, for a real potential the scheme of level lines here certainly transforms into itself under complex conjugation, i.e., under the reflection with respect to the real axis. (In the problem of underbarrier transition the potential is not obligatorily symmetric under the reflection $x \longrightarrow -x$.) And at last, here there is one more line of level — the real axis. The cuts starting at the turning points, $a$ and $a^*$, are conveniently drawn as presented in Fig. 5.7 (by analogy with Fig. 5.3).

Again, it is clear from physical arguments that on the real axis for $x \longrightarrow \infty$ the solution looks as $\psi_+^0$, where the index 0 means that its phase is counted off the origin. This solution should coincide with the wave $\psi_+^a$ on the level line 3a, which tends asymptotically to the real axis for $+\infty$. One should only include the factor related to the different starting points in $\psi_+^0$ and $\psi_+^a$:

$$\psi_+^0(z) \sim \exp\left\{ i \int_0^z k(z)dz \right\} = \exp\left\{ i \int_0^a k(z)dz + \int_a^z k(z)dz \right\},$$

so that

$$\psi_+^0(z) = r^{1/2}\psi_+^a(z),$$

$$r^{1/2} = \exp\left\{ i \int_0^a k(z)dz \right\} = \exp\left\{ -i \int_a^0 k(z)dz \right\} \sim \psi_-^a(0).$$

In this section we are not interested in the preexponential factor $k^{-1/2}$, and therefore we do not write it explicitly; of course, when constructing the correct solutions $\psi^a$, this factor is taken into account properly.

Since in the sector between the level lines 2a and 3a the solution $\psi_-^a$ is exponentially small, then

$$r^{1/2} = \exp\left\{-i\int_a^0 k(z)dz\right\} = \exp\left\{-\int_a^0 |k(z)|dz\right\} \ll 1\,.$$

The arguments, quite analogous to those used in the previous sections, allow one to construct easily the solution on the line 2a. It is equal to $r^{1/2}(\psi_+^a - i\psi_-^a)$, and should coincide with the asymptotic solution on the left real semi-axis. Now we go over in this solution to counting the phase from the origin, and thus rewrite it in the form $\psi_+^0 - ir\,\psi_-^0$. As a result, the correspondence between the asymptotic solutions on the left and right real semi-axes is

$$\psi_+^0 - ir\psi_-^0 \longrightarrow \psi_+^0\,.$$

This relation can also be improved since the sum of the reflection and transition coefficients, i.e., of $R = r^2$ and the modulus squared of the coefficient at $\psi_+^0$, respectively, should be equal to unity. As a result, we obtain

$$\psi_+^0 - i\,r\,\psi_-^0 \longrightarrow \left(1 - \frac{1}{2}r^2\right)\psi_+^0\,. \tag{5.12}$$

We note that the coefficient of the overbarrier reflection $R = r^2$ can be written as (compare with (5.9))

$$R = r^2 = \exp\left\{-2i\int_a^{a^*} k(z)dz\right\} = \exp\left\{-2\int_a^{a^*} |k(z)|dz\right\} \ll 1\,. \tag{5.13}$$

The quite natural question may arise: what will happen if the solution on the real axis will be constructed starting from the lines of level $1a^*$, $2a^*$, originating from the turning point $a^*$? The point is that in the region between the lines of level $1a^*$ and $2a^*$ the solution $\psi_+^{a^*}$ is exponentially small. Therefore, if this solution is valid on the line $1a^*$ which goes to the right, then it is the same on the line $2a^*$, going to the left. In other words, here we do not see any overbarrier reflection at all (as distinct from the transition from the line $3a$ to the line $2a$).

To see the reason for this discrepancy, let us express our solution (5.12) for $z \longrightarrow -\infty$ via the solutions counted off the turning point $a^*$:

$$\psi_+^0 - i\,r\,\psi_-^0 = r^{-1/2}\psi_+^{a^*} - i\,r^{3/2}\psi_-^{a^*} = r^{-1/2}[\psi_+^{a^*} - i\,r^2\psi_-^{a^*}]\,.$$

Thus, on the line of level $2a^*$ the coefficient at the reflected wave is of second order in the parameter $|r| \ll 1$, as compared to the coefficient at the incoming wave. However, effects of second order in the exponentially small quantity $r$ are beyond the accuracy of our calculation. In other words, there is no contradiction with the solution (5.12). However, the analysis of the lower system of levels simply does not guarantee the accuracy required for calculating the coefficient of overbarrier reflection in the used approach.

At last, we present in this case as well the result for the transmission and reflection coefficients, derived without the assumption that the semiclassical approximation is applicable between the turning points, i.e., without the assumption that $r^2$ is exponentially small:

$$D = \frac{1}{1+r^2}; \quad R = \frac{r^2}{1+r^2}; \quad D+R=1. \tag{5.14}$$

In conclusion, let us come back to the exponentially small result (5.13). Its applicability is in fact essentially limited for the following reason. This result is true under the assumption that the potential is described by a smooth analytic function. Under actual conditions, however, a real physical potential can contain random perturbations, of the characteristic scale comparable to the de Broglie wavelength or even less than it. Though being small, such irregular perturbation results typically in a contribution to the coefficient of the overbarrier reflection which falls down not exponentially with momentum, but much weaker, only as its power[6]. Whichever of the effects dominates in the overbarrier reflection, the regular exponentially small one (5.13), or the powerlike suppressed effect of small random perturbations, depends in fact on the concrete conditions of the problem.

## 5.6 What Is the Accuracy of Adiabatic Invariant?

The problem of the accuracy of adiabatic invariants is an example of application of differential equations with a small coefficient at the higher derivative beyond quantum mechanics.

As a concrete problem, let us consider a linear oscillator with a frequency dependent on time. The corresponding differential equation

$$\ddot{x} + \omega^2(t)x = 0$$

coincides, up to notations, with equation (5.2). As to the frequency $\omega(t)$, we will assume that it is a slowly varying function of $t$, which tends to the same limit $\omega$ for $t \longrightarrow \pm\infty$. (Correspondingly, in equation (5.2) $k(x)$ tends to $k = \sqrt{2mE}/\hbar$ for $x \longrightarrow \pm\infty$ where the potential $U(x)$ turns to zero.)

Following *A.M. Dykhne* (1960), we will use the fact that this problem, up to notations, coincides with the problem of the overbarrier reflection, considered above. One should only take into account that the physical coordinate $x_\pm$, as distinct from $\psi_\pm$, is real. Thus, the solutions at infinity look here as follows:

---

[6]See, for instance: L. D. Landau and E. M. Lifshitz, *Quantum Mechanics*. §52, Problem 3.

$$x(t \longrightarrow -\infty) = \mathrm{Re}(x_+ - ir\,x_-) = x_0\,(\cos\Phi - r\sin\Phi)\,,$$

$$x(t \longrightarrow \infty) = \mathrm{Re}\left(1 - \frac{r^2}{2}\right)x_+ = x_0\left(1 - \frac{r^2}{2}\right)\cos\Phi\,.$$

Here we include in $x_0$ the preexponential factor $w^{-1/2}(t)$; anyway, we are not going to differentiate it, since it would go beyond the accuracy of the employed semiclassical approximation.

We note now that for a harmonic oscillator the adiabatic invariant

$$I = (2\pi)^{-1} \oint p\,dq$$

is equal to the ratio of the Hamiltonian $H$ to the frequency $w^7$. Let us consider this ratio for $t \longrightarrow \pm\infty$ (recalling that $\dot{\Phi} = w$ by definition):

$$I(t \longrightarrow -\infty) = \frac{m}{2w}\,(\dot{x}^2 + w^2 x^2)|_{t \longrightarrow -\infty}$$

$$= \frac{mwx_0^2}{2}\left[(-\sin\Phi - r\,\cos\Phi)^2 + (\cos\Phi - r\sin\Phi)^2\right] = \frac{mwx_0^2}{2}\,(1 + r^2)\,,$$

$$I(t \longrightarrow \infty) = \frac{mwx_0^2}{2}\,(1 - r^2)\,.$$

Their difference,

$$I(t \longrightarrow \infty) - I(t \longrightarrow -\infty) = -m\,wx_0^2 r^2, \tag{5.15}$$

is exponentially small, together with $r^2$ (see (5.13)).

In other words, the adiabatic invariant of a harmonic oscillator is conserved with an exponential accuracy for slowly varying frequency[8].

---

[7]To convince oneself of this fact, it is sufficient for instance to compare in quantum mechanics the eigenvalues of the Hamiltonian and the truncated action (which is an adiabatic invariant), $\hbar w(n + 1/2)$ and $\hbar(n + 1/2)$, respectively.

[8]Of course, this conclusion is also valid only under the condition that the discussed regular exponentially small effect dominates over that of small random perturbations of the potential (see Section 5.5).

# 6

# Quantum Electrodynamics. Again Minimum Calculations

## 6.1 Spin in Electromagnetic Field

### 6.1.1 Covariant Equation of Motion of Spin

We start with the spin precession for a nonrelativistic charged particle. The equation that describes this precession is well-known:

$$\dot{\mathbf{s}} = \frac{eg}{2m} \left[ \mathbf{s} \times \mathbf{B} \right].$$

(6.1)

Here $\mathbf{B}$ is an external magnetic field, $e$ and $m$ are the charge and mass of the particle, $g$ is its gyromagnetic ratio (for electron $g \approx 2$)[1]. In other words, the spin precesses around the direction of magnetic field with the frequency $-(eg/2m)\mathbf{B}$. In the same nonrelativistic limit the velocity precesses around the direction of $\mathbf{B}$ with the frequency $-(e/m)\mathbf{B}$:

$$\dot{\mathbf{v}} = \frac{e}{m} \left[ \mathbf{v} \times \mathbf{B} \right].$$

(6.2)

Thus, for $g = 2$ spin and velocity precess with the same frequency, so that the angle between them is conserved.

Let us note that both equations, (6.1) and (6.2), hold as Heisenberg equations of motion in an external field for the spin and velocity operators, $\mathbf{s}$ and $\mathbf{v}$. On the other hand, being averaged over properly localized wave packets, these equations go over into (semi)classical equations of motion for spin and velocity. This refers also to the relativistic generalizations of equation (6.1), discussed in this section below.

We will consider at first the covariant semiclassical formalism using the four-dimensional vector of spin $S_\mu$. This 4-vector is defined as follows. In the particle rest frame $S_\mu$ has no time component and reduces to the common three-dimensional vector of spin $\mathbf{s}$, i.e., in this frame $S_\mu = (0, \mathbf{s})$. In the

---

[1]We put the velocity of light $c$ equal to 1.

reference frame where the particle moves with velocity $\mathbf{v}$, the vector $S_\mu$ is constructed from $(0, \mathbf{s})$ by means of the Lorentz transformation, so that here

$$S_0 = \gamma \mathbf{v s}, \quad \mathbf{S} = \mathbf{s} + \frac{\gamma^2 \mathbf{v}(\mathbf{v s})}{\gamma + 1}. \tag{6.3}$$

Then, just by definition of $S_\mu$, the following identities take place:

$$S_\mu S_\mu = -\mathbf{s}^2 \, (= \text{const}), \quad S_\mu u_\mu = 0\,; \tag{6.4}$$

as usual, here $u_\mu$ is the four-velocity.

The right-hand side of the equation for $dS_\mu/d\tau$ should be linear and homogeneous both in the electromagnetic field strength $F_{\mu\nu}$, and in the same four-vector $S_\mu$, and may depend also on $u_\mu$. By virtue of the first identity (6.4), the right-hand side should be four-dimensionally orthogonal to $S_\mu$. Therefore, the general structure of the equation we are looking for is

$$\frac{dS_\mu}{d\tau} = \alpha F_{\mu\nu} S_\nu + \beta u_\mu F_{\nu\lambda} u_\nu S_\lambda. \tag{6.5}$$

Comparing the nonrelativistic limit of this equation with (6.1), we find

$$\alpha = \frac{eg}{2m}.$$

Now we take into account the second identity (6.4), which after differentiation in $\tau$ gives

$$u_\mu \frac{dS_\mu}{d\tau} = -S_\mu \frac{du_\mu}{d\tau},$$

and recall the classical equation of motion for a charge:

$$m \frac{du_\mu}{d\tau} = e F_{\mu\nu} u_\nu. \tag{6.6}$$

Then, multiplying equation (6.5) by $u_\mu$, we obtain

$$\beta = -\frac{e}{2m}(g - 2).$$

Thus, the covariant equation of motion for spin is

$$\frac{dS_\mu}{d\tau} = \frac{eg}{2m} F_{\mu\nu} S_\nu - \frac{e}{2m}(g - 2) u_\mu F_{\nu\lambda} u_\nu S_\lambda \tag{6.7}$$

(*Ya. I. Frenkel*, 1926; *L. Thomas*, 1927; *V. Bargman, L. Michel, V. Telegdi*, 1959).

Let us discuss the limits of applicability for this equation.

Of course, typical distances at which the trajectory changes (for instance, the Larmor radius in a magnetic field) should be large as compared to the de Broglie wavelength $\hbar/p$ of the elementary particle. Then, the external field itself should not change essentially at the distances on the order of the de Broglie wavelength $\hbar/p$. For ultrarelativistic particles, when the Compton wavelength $\hbar/(mc)$ exceeds the de Broglie one, the external field should not change essentially at the distances $\sim \hbar/(mc)$. If the last condition does not hold, the scatter of velocities in the rest frame is not small as compared to $c$, and one cannot use in this frame the nonrelativistic formulas.

Besides, if the external field changes rapidly, the motion of spin will be influenced by interaction of higher electromagnetic multipoles of the particle with field gradients. For a particle of spin $1/2$ higher multipoles are absent, and the gradient-dependent effects are due to finite form factors of the particle. These effects start here at least in second order in field gradients and usually are negligible.

At last, in equation (6.7) we confine ourselves to effects of first order in the external field. This approximation relies in fact on the implicit assumption that the first-order interaction with the external field is less than the excitation energy of the spinning system. Usually this assumption is true and the first-order equation (6.7) is valid. Still, one can easily point out situations when this is not the case. To be definite, let us consider the hydrogen-like ion $^3\text{He}^+$ in the ground $s$-state with the total spin $F = 1$. It can be easily demonstrated that an already quite moderate external magnetic field is sufficient to break the hyperfine interaction between the electron and nuclear magnetic moments (a sort of Paschen–Back effect). Then, instead of a precession of the total spin $\mathbf{F}$ of the ion, which should be described by equations (6.1) or (6.7) with a corresponding ion $g$-factor, we will have a separate precession of the decoupled electron and nuclear spins.

Let us go back now to equation (6.7). We note that for $g = 2$ and in the absence of an electric field, its zeroth component reduces to

$$\frac{dS_0}{d\tau} = 0.$$

Taking into account definition (6.3) for $S_0$ and the fact that in a magnetic field the particle energy remains constant, we find immediately that the projection of spin $\mathbf{s}$ onto velocity, so-called helicity, is conserved.

### 6.1.2 Noncovariant Equation of Motion for Spin of Relativistic Particle. Thomas Precession

We will obtain now the relativistic equation for the three-dimensional vector of spin $\mathbf{s}$, that directly describes the internal angular momentum of a particle in its "momentary" rest frame. This equation can be derived from (6.7) using relations (6.3), together with the equations of motion for a charge in external field. It will require, however, quite tedious calculations. Therefore, we choose another way, somewhat simpler and much more instructive.

First, we transform equation (6.1) from the comoving inertial frame, where the particle is at rest, into the laboratory one. The magnetic field $\mathbf{B}'$ in the rest frame is expressed via the electric and magnetic fields **E** and **B** given in the laboratory frame, as follows:

$$\mathbf{B}' = \gamma \mathbf{B} - \frac{\gamma^2}{\gamma+1} \mathbf{v}(\mathbf{vB}) - \gamma \mathbf{v} \times \mathbf{E}.$$

This expression can be easily checked by comparing it component by component with the transformation of magnetic field for two cases: when this field is parallel to the velocity and orthogonal to it, respectively. Then one should take into account that the frequency in the rest-frame time $t$ is $\gamma$ times smaller than the frequency in the laboratory time $\tau$ (indeed, $d/dt = d\tau/dt \cdot d/d\tau = \gamma^{-1}d/d\tau$). Found in this way the contribution to the precession frequency is

$$\boldsymbol{\omega}_g = -\frac{eg}{2m}\left[\mathbf{B} - \frac{\gamma}{\gamma+1}\mathbf{v}(\mathbf{vB}) - \mathbf{v} \times \mathbf{E}\right].$$

However, it is clear from equation (6.7) that spin precesses even if $g = 0$. To elucidate the origin of this effect, the so-called Thomas precession (*L. Thomas*, 1926), we consider two successive Lorentz transformations: at first from the laboratory frame $S$ into the frame $S'$ that moves with the velocity **v** with respect to $S$, and then from $S'$ into the frame $S''$ that moves with respect to $S'$ with the infinitesimal velocity $d\mathbf{v}$. Let us recall in this connection the following fact related to usual three-dimensional rotations: the result of two successive rotations with respect to noncollinear axes $\mathbf{n}_1$ and $\mathbf{n}_2$ contains in particular a rotation around the axis directed along their vector product $\mathbf{n}_1 \times \mathbf{n}_2$. Now it is only natural to assume that the result of the above successive Lorentz transformations will contain in particular a usual rotation around the axis directed along $d\mathbf{v} \times \mathbf{v}$. In the result, spin in the rest frame will rotate in the opposite direction by an angle which we denote by $\varkappa[d\mathbf{v} \times \mathbf{v}]$. Here $\varkappa$ is some numerical factor to be determined below. It depends generally speaking on the particle energy.

This is in fact the Thomas precession. Its frequency in the proper time $\tau$ is

$$\boldsymbol{\omega}'_T = \varkappa[d\mathbf{v}/d\tau \times \mathbf{v}] = \varkappa\frac{e}{m}[\mathbf{E}' \times \mathbf{v}].$$

Now we transform the electric field $\mathbf{E}'$ from the proper frame into the laboratory one, as was done above for the magnetic field $\mathbf{B}'$, and go over also from the proper time $\tau$ to $t$. In the result, the frequency of the Thomas precession in the laboratory frame is

$$\boldsymbol{\omega}_T = \varkappa\frac{e}{m}\left[\left(\mathbf{E} - \frac{\gamma}{\gamma+1}\mathbf{v}(\mathbf{vE}) + \mathbf{v} \times \mathbf{B}\right) \times \mathbf{v}\right]$$

$$= -\varkappa\frac{e}{m}\left[\mathbf{v} \times \mathbf{E} - v^2\mathbf{B} + \mathbf{v}(\mathbf{vB})\right].$$

To find the coefficient $\varkappa$, we recall that in a magnetic field, for $g = 2$ the projection of spin onto the velocity is conserved. In other words, in this case the total frequency of the spin precession $\omega = \omega_g + \omega_T$ coincides with the frequency of the velocity precession which is well known to be

$$\omega_v = - \frac{e}{m\gamma} \mathbf{B}.$$

From this we find easily that $\varkappa = \gamma/(\gamma + 1)$. Correspondingly, the relativistic equation of motion for the three-dimensional vector of spin $\mathbf{s}$ in external electromagnetic field is

$$\frac{d\mathbf{s}}{dt} = (\omega_g + \omega_T) \times \mathbf{s} = \frac{e}{2m} \left\{ \left( g - 2 + \frac{2}{\gamma} \right) [\mathbf{s} \times \mathbf{B}] \right.$$

$$\left. - (g - 2) \frac{\gamma}{\gamma + 1} [\mathbf{s} \times \mathbf{v}](\mathbf{vB}) - \left( g - \frac{2\gamma}{\gamma + 1} \right) [\mathbf{s} \times [\mathbf{v} \times \mathbf{E}]] \right\}. \tag{6.8}$$

We note also that the relativistic Hamiltonian for the interaction of the three-dimensional vector of spin with external electromagnetic field is written in the usual form:

$$H = \boldsymbol{\omega}\mathbf{s}. \tag{6.9}$$

Not only does it generate via the standard relation

$$\frac{d\mathbf{s}}{dt} = \frac{i}{\hbar} [H, \mathbf{s}] \tag{6.10}$$

equation (6.8). For instance, it is a simple problem to obtain with this Hamiltonian equations of motion of the quadrupole moment of a relativistic particle in electric and magnetic fields, neglecting the field gradients. In the particle rest frame, the operator of its quadrupole moment is

$$q_{mn} = \frac{3q}{2s(2s - 1)} \left[ s_m s_n + s_n s_m - \frac{2}{3} s(s + 1)\delta_{mn} \right];$$

here the structure in square brackets guarantees the symmetry and vanishing trace of this operator, $q_{mn} = q_{nm}$, $q_{mm} = 0$; the overall factor at the square brackets corresponds to the normalization condition $q_{zz} = q$ for $s_z = s$. To calculate the commutator in the corresponding equation

$$\frac{dq_{mn}}{dt} = \frac{i}{\hbar} [\omega_k s_k, q_{mn}], \tag{6.11}$$

is an elementary problem.

## 6.2 Spin in Electromagnetic Field. New Applications

### 6.2.1 Single-Photon Radiative Transition between Atomic $s$-Levels

The discussed transition is a magnetic dipole one since the initial and final states have the same parity. However, this $M1$ transition is strongly forbidden. Indeed, the matrix element of the magnetic dipole moment operator

$\mu = (e/2m)(1+\sigma)$ (here and in the next subsection we put $\hbar = 1$) vanishes in this case since the radial wave functions of two different atomic states are orthogonal. Besides, the matrix element of the orbital momentum operator $\mathbf{l}$ between $s$-states turns to zero identically.

As to the operator $\sigma$, its contribution does not vanish due to the retardation effects in the magnetic field, i.e., due to the higher terms of the expansion in the ratio of the atomic size to the wavelength:

$$< 2|(\sigma \mathbf{B})\exp(i\mathbf{kr})|1 >\approx -\frac{1}{6}(\sigma \mathbf{B})k^2 < 2|r^2|1 >; \qquad (6.12)$$

here we have taken into account the spherical symmetry of the wave functions. The expression $k^2 < 2|r^2|1 >$ transforms as follows:

$$k^2 < 2|r^2|1 >=< 2|(E_2 - E_1)^2 r^2|1 >=< 2|[H_0[H_0, r^2]]|1 >$$

$$= \frac{2}{m} < 2|2U + r(dU/dr)|1 >, \quad H_0 = \frac{p^2}{2m} + U(r). \qquad (6.13)$$

In line with this, essentially relativistic effect of second order in $v/c$, one should take into account the corrections of the same order in the Hamiltonian (6.9) for the electron. The corresponding approximate Hamiltonian is conveniently presented as follows (for the time being, for an arbitrary spin and $g$-factor):

$$H' = -\frac{e}{2m}\left\{(g-v^2)\mathbf{B} - \frac{1}{2}(g-2)\mathbf{v}(\mathbf{v}B) - (g-1)\mathbf{v}\times\mathbf{E}\right\}\mathbf{s}$$

$$= -\frac{e}{2m}\left\{\left(g - \frac{p^2}{m^2}\right)\mathbf{B} - \frac{(g-2)}{2m^2}\mathbf{p}(\mathbf{p}B)\right. \qquad (6.14)$$

$$\left. -\frac{(g-1)}{m}(\mathbf{p} - e\mathbf{A})\times\mathbf{E}\right\}\mathbf{s}.$$

In the case under discussion now, that of an atomic electron, one can put with good accuracy $g = 2$ and neglect the second term in the curly brackets in formula (6.14). Besides, we go over in this formula from the spin operator $\mathbf{s}$ to $\sigma = 2\mathbf{s}$. As a result, we obtain

$$H' = -\frac{e}{2m}\left\{\left(1 - \frac{p^2}{2m^2}\right)\mathbf{B} - \frac{1}{2m}(\mathbf{p} - e\mathbf{A})\times\mathbf{E}\right\}\sigma. \qquad (6.15)$$

As to the correction $-(p^2/2m^2)\mathbf{B}$, the corresponding matrix element is

$$-\frac{1}{m}\mathbf{B} < 2|p^2/2m|1 >= \frac{1}{m}\mathbf{B} < 2|U(r)|1 > . \qquad (6.16)$$

The vector product in the curly brackets is rewritten as follows:

$$-\frac{1}{2m}(\mathbf{p} - e\mathbf{A})\times\mathbf{E} \longrightarrow -\frac{1}{2m}[\mathbf{A}\times\nabla U] - \frac{1}{4m}[\mathbf{p}\times\mathbf{E} - \mathbf{E}\times\mathbf{p}];$$

we take into account below the dependence of $\mathbf{E}$ on $\mathbf{r}$, therefore the second structure here is written in an explicitly Hermitian form. When calculating the matrix element $< \mathbf{A} \times \nabla U >$, one can neglect the dependence of the photon magnetic field on coordinates. Then, the calculation is simplified by choosing for the field $\mathbf{A}$ of the wave the gauge where $\mathbf{A} = (1/2)\mathbf{B} \times \mathbf{r}$:

$$-\frac{1}{2m} < 2| \frac{1}{2} [\mathbf{B} \times \mathbf{r}] \times (\mathbf{r}/r)(dU/dr)|1 >= \frac{\mathbf{B}}{6m} < 2| r(dU/dr)|1 > . \quad (6.17)$$

Finally, when calculating the last term, the electric field of the wave is needed to first order in $\omega$. In the same gauge, one can put here $\mathbf{E} = (i\omega/2)\,\mathbf{B} \times \mathbf{r}$. Then

$$\frac{i\omega}{8m} < 2| [\mathbf{B} \times \mathbf{r}] \times \mathbf{p} - \mathbf{p} \times [\mathbf{B} \times \mathbf{r}]|1 >$$

$$= -\frac{i\mathbf{B}}{12m} < 2| [H_0, \mathbf{rp} + \mathbf{pr}]|1 >= \frac{\mathbf{B}}{6m} < 2| 2U + r(dU/dr)|1 > . \quad (6.18)$$

With the account for (6.13) – (6.18), the matrix element of Hamiltonian (6.15) reduces to the compact form:

$$< 2|H'|1 >= -\frac{e}{m} \sigma \mathbf{B} \frac{1}{3m} < 2|U|1 > .$$

Thus, the effective operator of the electromagnetic transition between $s$-levels can be presented as

$$H_{\text{eff}} = -\frac{e}{2m} \sigma \mathbf{B} \frac{2}{3} \frac{U}{m} . \quad (6.19)$$

This operator applies not only to atomic hydrogen and hydrogen-like ions. For instance, it applies also to the outer electron in alkaline atoms, of course as long as one can confine to the single-particle approximation.

Quite standard calculation with operator (6.19) leads to the following result for the probability of transition $2s \rightarrow 1s + \gamma$ in hydrogen (where $U = -\alpha/r$):

$$W_\gamma = \frac{\alpha^{11}m}{2^2 3^5} = 0.25 \times 10^{-5}\,\text{s}^{-1}. \quad (6.20)$$

In other words, the lifetime of the $2s$-state with respect to the discussed single-photon transition is 4.7 days.

In fact, the main channel of the decay of the $2s$-state in the absence of perturbations (external electric field, collisions) is the transition $2s \rightarrow 1s + 2\gamma$ with the emission of two $E1$-photons. Its probability is $W_{2\gamma} = 8.23\,\text{s}^{-1}$.

## 6.2.2 Low-Energy Theorems for Compton Scattering

We start with the scattering of an electromagnetic wave on a charged spinless particle. It is convenient to use here the Coulomb gauge where the vector potential of the wave satisfies the conditions $A_0 = 0$ and $\nabla \mathbf{A} = 0$ (or $\mathbf{kA} = 0$

in the momentum representation). The structure of the operator of first order in the field is

$$V_1 \sim -e\,(\mathbf{p}' + \mathbf{p})\,\mathbf{A} \tag{6.21}$$

both in the nonrelativistic problem and in the relativistic one. We will use the coordinate frame where the scatterer is initially at rest. It can be easily seen that in this frame the operator $V_1$ reduces to $\mathbf{kA}$ either for the incoming photon (see Fig. 6.1, $a$), or for the scattered one (see Fig. 6.1, $b$) and thus turns to zero. Together with it, the contribution to the Compton amplitude arising in second order in the operator $V_1$ vanishes as well.

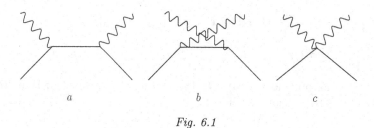

$$a \qquad\qquad\qquad b \qquad\qquad\qquad c$$

*Fig. 6.1*

Let us go over now to the contribution of the operator of second order in the field (see Fig. 6.1, $c$)

$$V_2 \sim e^2\,\mathbf{A}^2\,. \tag{6.22}$$

If the frequency of the incoming photon is much smaller than the mass $m$ of the target, then the velocity of the target after scattering is small, so that the interaction of second order in the low-frequency limit is described by the usual nonrelativistic operator

$$V = \frac{e^2}{2m}\,\mathbf{A}^2\,. \tag{6.23}$$

This operator generates the so-called Thomson amplitude

$$A_{\text{Th}} = -\frac{e^2}{m}\,(\mathbf{e}_2^*\mathbf{e}_1)\,. \tag{6.24}$$

Here $\mathbf{e}_{1,2}$ are the polarization vectors of the photons, incoming and scattered, correspondingly (in this subsection we omit the standard normalization factors $\sqrt{4\pi/2\omega}$ at them). The overall numerical factor $1/2$ in operator (6.23) has disappeared from this amplitude, in accordance with two options for identifying the photons, initial and final, with two operators $\mathbf{A}$ in (6.23). We use here and below the standard definition of the scattering amplitude where its sign is opposite to that of the effective interaction.

We have derived here the Thomson amplitude (6.24) in fact under the assumption that the particle interacts with the electromagnetic field to the lowest nonvanishing order of perturbation theory. As to other possible interactions of this particle, they were all neglected. Therefore, the charge and mass in this amplitude initially refer to a free particle, i.e., they are the bare, unrenormalized ones. What will change if we take into account other possible interactions of the scatterer, including also that with virtual photons? Certainly, these interactions will change the bare values of the charge and mass, i.e., will renormalize them. Therefore in fact one should ascribe to $e$ and $m$ in the amplitude (6.24) their observable, physical values.

Then, the problem is whether the interactions can change besides the numerical factor in expression (6.24). In the employed approximation $\omega \ll m$ this factor (equal to unity) coincides with the properly normalized time component of the current density (in the momentum representation) at the vanishing momentum transfer. And the last quantity is nothing but the total number of particles, which certainly does not change when interactions are switched on.

Thus, formula (6.24) is the exact expression for the scattering amplitude in the low-frequency limit, and $e$ and $m$ in it are the observable, physical charge and mass of the scatterer. This is the first low-energy theorem for the Compton scattering.

Let us go over now to the second low-energy theorem related to the spin of the scatterer. It is quite natural that when the particle spin and its precession in the field of the wave are included in consideration, a new contribution to the scattering amplitude will arise. To calculate this effect, we use Hamiltonian (6.14). In its terms linear in the field, we confine to the contribution of first order in $1/c$, and of course keep the term quadratic in the field (and in $1/c$):

$$H'' = -\frac{eg}{2m}\,\mathbf{B}\mathbf{s} - \frac{e^2(g-1)}{2m^2}\,[\mathbf{A} \times \mathrm{e}]\,\mathbf{s}. \tag{6.25}$$

The diagrams which describe the discussed effect are the same as those presented in Fig. 6.1.

We start with the contribution of diagrams 6.1, $a$ and 6.1, $b$, which are generated by the first term in Hamiltonian (6.25). Let us note here that in the noncovariant perturbation theory, employed here, the energy of the initial state, counted off $m$, is $\omega$ in both diagrams, and the intermediate state energies are 0 and $2\omega$, respectively. The total contribution of these two diagrams to the scattering amplitude is

$$A_1 = -\frac{e^2 g^2}{4m^2}\left\{ \frac{([\mathbf{k}_2 \times \mathbf{e}_2^*]\mathbf{s})([\mathbf{k}_1 \times \mathbf{e}_1]\mathbf{s})}{\omega} + \frac{([\mathbf{k}_1 \times \mathbf{e}_1]\mathbf{s})([\mathbf{k}_2 \times \mathbf{e}_2^*]\mathbf{s})}{-\omega} \right\}$$

$$= -\frac{e^2 g^2}{4m^2}\,i\,\omega\,[[\mathbf{n}_2 \times \mathbf{e}_2^*] \times [\mathbf{n}_1 \times \mathbf{e}_1]]\,\mathbf{s}; \tag{6.26}$$

here and below $\mathbf{n}_{1,2} = \mathbf{k}_{1,2}/k = \mathbf{k}_{1,2}/\omega$.

The contribution to the scattering amplitude of diagram 6.1, $c$, which is generated here by the second term in Hamiltonian (6.25), is

$$A_2 = \frac{e^2(g-1)}{m^2} \, i\omega \, [\mathbf{e}_2^* \times \mathbf{e}_1] \, \mathbf{s} \, . \tag{6.27}$$

And at last, the third contribution to the spin-dependent amplitude arises also from diagrams 6.1, $a$, and 6.1, $b$, in each of which now the left vertex is generated by the first term in Hamiltonian (6.25), and the right one by the common nonrelativistic interaction

$$-\frac{e}{2m} \, (\mathbf{p} + \mathbf{p}')\mathbf{A} \, .$$

This contribution is

$$A_3 = -\frac{e^2 g}{4m^2} \, i\omega \, [(\mathbf{n}_1 \mathbf{e}_2^*)[\mathbf{n}_1 \times \mathbf{e}_1] - (\mathbf{n}_2 \mathbf{e}_1)[\mathbf{n}_2 \times \mathbf{e}_2^*]] \, \mathbf{s} \, . \tag{6.28}$$

The total low-energy spin-dependent amplitude is written as follows:

$$A = A_1 + A_2 + A_3 = -\frac{e^2}{4m^2} \, i\omega \, \{(g-2)^2 \, [\mathbf{e}_2^* \times \mathbf{e}_1]$$

$$-g^2 \, [[1 - (\mathbf{n}_1 \mathbf{n}_2)][\mathbf{e}_2^* \times \mathbf{e}_1] + [\mathbf{n}_1 \times \mathbf{n}_2](\mathbf{e}_2^* \mathbf{e}_1)$$

$$-[\mathbf{n}_1 \times \mathbf{e}_2^*](\mathbf{n}_2 \mathbf{e}_1) + (\mathbf{n}_1 \mathbf{e}_2^*)[\mathbf{n}_2 \times \mathbf{e}_1]]$$

$$-g \, [[\mathbf{n}_2 \times \mathbf{e}_2^*](\mathbf{n}_2 \mathbf{e}_1) - (\mathbf{n}_1 \mathbf{e}_2^*)[\mathbf{n}_1 \times \mathbf{e}_1]]\} \, \mathbf{s} \, . \tag{6.29}$$

To derive this formula, expression (6.26) was transformed by means of identity

$$[\mathbf{n}_2 \times \mathbf{e}_2^*] \times [\mathbf{n}_1 \times \mathbf{e}_1] = (\mathbf{n}_1 \mathbf{n}_2)[\mathbf{e}_2^* \times \mathbf{e}_1] - [\mathbf{n}_1 \times \mathbf{n}_2](\mathbf{e}_2^* \mathbf{e}_1)$$

$$+ [\mathbf{n}_1 \times \mathbf{e}_2^*](\mathbf{n}_2 \mathbf{e}_1) - (\mathbf{n}_1 \mathbf{e}_2^*)[\mathbf{n}_2 \times \mathbf{e}_1] \, . \tag{6.30}$$

As well as the Thomson amplitude (6.24), result (6.29) was obtained under the assumption that the particle interacts with the electromagnetic field to the lowest nonvanishing order of perturbation theory, and other possible interactions of this particle were all neglected. In the present case as well the role of all interactions, unaccounted for (electromagnetic radiative corrections included), reduces to the fact that the mass and the charge, $e$ and $m$, in amplitude (6.29) are the observable physical ones. Certainly, for the $g$-factor one should use its true observable value as well. As to the overall numerical factor in expression (6.29), it does not change here by virtue of the conservation law for the angular momentum of the scatterer in its rest frame, i.e., for the spin $s$.

Thus, formula (6.29) is the exact expression for the scattering amplitude in the low-frequency limit, and $e$, $m$, and $g$ in this formula are, respectively, the observable physical charge, mass, and gyromagnetic ratio of the target. This is the second low-energy theorem for the Compton scattering.

Result (6.29) simplifies essentially for the forward scattering when $\mathbf{n}_1 = \mathbf{n}_2 = \mathbf{n}$:

$$A_n = A_1 + A_2 = -\frac{e^2(g-2)^2}{4m^2}\, i\omega \,[\mathbf{e}_2^* \times \mathbf{e}_1]\,\mathbf{s}\,. \tag{6.31}$$

This amplitude turns to zero for $g = 2$.

Relations (6.29) and (6.31) were derived by *F.E. Low* (1954), and by *M. Gell-Mann, M.L. Goldberger* (1954).

One more case when result (6.29) simplifies essentially is the scattering without change of polarization. Here

$$\mathbf{e}_2 = \mathbf{e}_1 = \mathbf{e}, \quad (\mathbf{n}_{1,2}\,\mathbf{e}) = (\mathbf{n}_{1,2}\,\mathbf{e}^*) = 0\,,$$

so that the amplitude reduces to

$$A_e = -\frac{e^2}{4m^2}\, i\omega \,\big\{(g-2)^2\,[\mathbf{e}^* \times \mathbf{e}]$$
$$-g^2\,[\,[1-(\mathbf{n}_1\mathbf{n}_2)][\mathbf{e}^* \times \mathbf{e}] + [\mathbf{n}_1 \times \mathbf{n}_2]\,\xi\,]\big\}\,\mathbf{s}\,, \tag{6.32}$$

where $\xi = |\mathbf{e}|^2 \leq 1$ is the degree of polarization of the photon.

Let us confine ourselves here as well to the forward scattering. Then

$$-i[\mathbf{e}^* \times \mathbf{e}] = \xi_c \mathbf{n}\,, \tag{6.33}$$

where $\xi_c$ is the degree of circular polarization, and amplitude (6.32) looks as follows:

$$A_{en} = \frac{e^2\,(g-2)^2}{4m^2}\,\omega\,\xi_c(\mathbf{n}\,\mathbf{s})\,. \tag{6.34}$$

We recall now that the refraction index $n$ for a medium with the density $N$ of scatterers is expressed via the forward scattering amplitude $A$ as follows:

$$n = 1 + \frac{2\pi N A}{\omega^2}\,.$$

Therefore, a medium consisting of polarized scatterers possesses by virtue of equation (6.34) a specific anisotropic optical activity. The corresponding correction to the refraction index of circularly polarized photon is

$$\Delta n = \pm\frac{\pi N e^2(g-2)^2 s}{2m^2\omega}\,\cos\theta\,; \tag{6.35}$$

here $\pm$ refers to the sign of the circular polarization, $\theta$ is the angle between the directions of the polarization of medium and of the propagation of photon; all the scatterers are assumed to be completely polarized.

## 6.3 Particle Production by Constant Electric Field

Particle production by an external electric field is a remarkable prediction of the relativistic quantum theory (*F. Sauter*, 1931; *J. Schwinger*, 1951). Electric field strengths necessary for the real observation of the effect are achieved in the collisions of highly charged atomic nuclei when they reach short distances.

Here, we will consider the model of this phenomenon which has an exact solution: the case of a constant homogeneous external field $\mathcal{E}$. We use the picture of the Dirac sea which simplifies the solution essentially.

Let us start with the calculation of the main, exponential dependence of the effect. We direct the $z$ axis along the constant force $\mathbf{F} = e\mathcal{E} = (0, 0, e\mathcal{E})$, then the potential energy is $U = -e\mathcal{E}z$. When a particle moves in such a field, both its total energy $E = \pm\sqrt{m^2c^4 + \mathbf{p}^2c^2} - e\mathcal{E}z$ and transverse momentum $\mathbf{p}_\perp = (p_x, p_y, 0)$ are conserved.

In this field the usual Dirac gap (Fig. 6.2, $a$) tilts (see Fig. 6.2, $b$). As a result, an electron with a negative energy in the absence of the field, can now tunnel through the gap (see the horizontal dashed line in Fig. 6.2, $b$) and go to infinity as a usual particle. The hole created in this way is nothing but a positron.

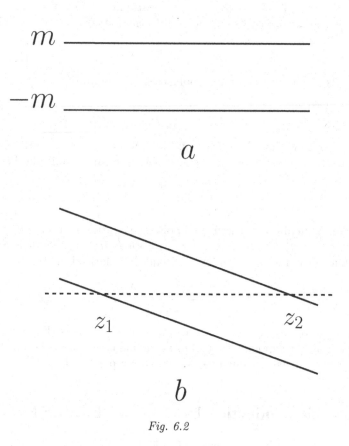

*Fig. 6.2*

Let $E = -\sqrt{m^2c^4 + \mathbf{p}^2c^2} - e\mathcal{E}z$ be the energy of a particle in the Dirac sea. The longitudinal momentum of the particle

$$p_z(z) = \frac{1}{c}\sqrt{(e\mathcal{E}z + E)^2 - m^2c^4 - \mathbf{p}_\perp^2 c^2}$$

vanishes at

$$z_{1,2} = \frac{-E}{e\mathcal{E}} \mp \frac{\sqrt{m^2c^4 + \mathbf{p}_\perp^2 c^2}}{e\mathcal{E}}.$$

An initial particle from the Dirac sea enters the barrier at the point $z = z_1$ and leaves it at $z = z_2$. The underbarrier action is found easily:

$$S = \int_{z_1}^{z_2} |p(z)|\,dz = \frac{\pi}{2}\frac{(m^2c^2 + \mathbf{p}_\perp^2)c}{e\mathcal{E}}.$$

As a result, the exponential factor in the probability $W$ of the underbarrier transition is

$$W \sim e^{-2S/\hbar} = \exp\left[-\frac{\pi(m^2c^2 + \mathbf{p}_\perp^2)c}{e\mathcal{E}\hbar}\right]. \tag{6.36}$$

Let us note that the external field can be considered constant if it changes weakly on the underbarrier path. The ratio $l/\lambda$ of the length of this path $l = z_2 - z_1 \sim mc^2/(e\mathcal{E})$ to the Compton wavelength of electron $\lambda = \hbar/(mc)$ is equal in the order of magnitude to the underbarrier action $S$ in the units of $\hbar$, so that in the semiclassical situation $l \gg \lambda$.

We calculate now the preexponential factor in the probability of pair creation [2]. The exponential (6.36) is the probability that a particle of the Dirac sea approaching the potential barrier from the left (see Fig. 6.2, b) will tunnel through it to the right, thus becoming a real electron. Let us consider the initial particles in the element of momentum space $d\mathbf{p} = d^2p_\perp\,dp_z$. Their space density is $dn = 2d\mathbf{p}/(2\pi\hbar)^3$, where the factor 2 corresponds to two possible orientations of the electron spin. The number of particles passing through the elementary area $dx\,dy$ on the left-hand side of the barrier is $d\dot{N} = d\,j_z(z)\,dx\,dy$, where $d\,j_z(z) = v_z(z)\,dn$. This expression contains the quantity

$$v_z(z)\,dp_z = \frac{\partial E}{\partial p_z}\,dp_z = dE,$$

where the partial derivative is taken at fixed values of $z$ and $\mathbf{p}_\perp$. On the other hand, it can be easily seen that the energy interval $dE$ of the tunneling particles is directly related to the interval $dz$ of the longitudinal coordinates of the points where particles enter the barrier: $dE = e\mathcal{E}\,dz$ (up to the sign, which is inessential here). To obtain the total number of pairs created in the volume $dV = dx\,dy\,dz$, we should multiply exponential (6.36) by $d\dot{N}$. As a

---

[2]The calculation presented below was performed for the first time by N.B. Narozhny (1979).

result, the total number of pairs created in the unit time in the unit volume is

$$P_{1/2} = \frac{dW}{dt\,dV} = 2\,e\mathcal{E} \int \frac{d^2 p_\perp}{(2\pi\hbar)^3} \exp\left[ - \frac{\pi(m^2 c^2 + \mathbf{p}_\perp^2)\,c}{e\mathcal{E}\hbar} \right].$$

Integrating this expression over the transverse momenta, we arrive at the final result:

$$P_{1/2} = \frac{e^2 \mathcal{E}^2}{4\pi^3 \hbar^2 c} \exp\left( - \frac{\pi m^2 c^3}{e\mathcal{E}\hbar} \right). \tag{6.37}$$

We have equipped the probability $W$ in the above formulas with the subscript $1/2$ to indicate that the result refers to particles of spin one-half. Obviously, the notion of the Dirac sea, and hence the above derivation by itself, do not apply to boson-pair creation. However, in the semiclassical approximation, the creation rate for particles of various spins differ by the number of spin states only. Thus, the probability of scalar particle production calculated in this approximation, is two times smaller:

$$P_0 = \frac{e^2 \mathcal{E}^2}{8\pi^3 \hbar^2 c} \exp\left( - \frac{\pi m^2 c^3}{e\mathcal{E}\hbar} \right). \tag{6.38}$$

The corresponding exact results for a constant electric field are:

$$P_{1/2} = \frac{e^2 \mathcal{E}^2}{4\pi^3 \hbar^2 c} \sum_{n=1}^{\infty} \frac{1}{n^2} \exp\left( -n \frac{\pi m^2 c^3}{e\mathcal{E}\hbar} \right),$$

$$P_0 = \frac{e^2 \mathcal{E}^2}{8\pi^3 \hbar^2 c} \sum_{n=1}^{\infty} \frac{(-1)^{n-1}}{n^2} \exp\left( -n \frac{\pi m^2 c^3}{e\mathcal{E}\hbar} \right).$$

Of course, the account for higher terms, with $n \geq 2$, in these sums makes sense only for very strong electric fields, for $\mathcal{E} \gtrsim (m^2 c^3/e\hbar)$. For smaller fields, simple formulas (6.37) and (6.38) are valid quantitatively.

In conclusion of this subsection, let us come back to the criterion of the semiclassical approximation, $l/\lambda \gg 1$. It means also that the tilt of the Dirac gap is very small. Therefore, the vicinity of the turning point, where the classical picture is inapplicable, extends anomalously far away into the (formally) classically accessible region. That is why the formation length for the electron – positron pairs is in this case not $(m/eE)$, as one may expect naïvely, but much larger, $(m/eE)(m^2/eE)^{1/2}$, as was demonstrated by direct calculations (*A.I. Nikishov*, 1969).

## 6.4 Vacuum Fluctuations of Electromagnetic Field and Shift of Levels

### 6.4.1 Lamb Shift for Electron in Coulomb Field

The shift of levels due to the quantum fluctuations of the vacuum of electromagnetic field, the so-called Lamb shift, is considered usually in textbooks using the whole machinery of quantum electrodynamics, even in a relatively simple problem of an electron in a Coulomb field.

The more intuitive approach, presented below, allows one to derive in a rather simple way leading, logarithmic terms in the answers not only for this problem, but for more complicated problems as well. We use here the Coulomb gauge, $\nabla \mathbf{A} = 0$, where the electromagnetic field is split into the instantaneous Coulomb field and three-dimensionally transverse photons.

We start with the application of this approach to the mentioned simple problem — electron in the Coulomb field (T. Welton, 1948). Due to the quantum fluctuations of the vacuum of electromagnetic field, the radius vector of a charged particle fluctuates as well: $\mathbf{r} \rightarrow \mathbf{r} + \boldsymbol{\rho}$. After averaging over these fluctuations, the potential energy is modified as follows:

$$< V(\mathbf{r} + \boldsymbol{\rho}) >= V(r) + < \rho_i > \nabla_i V(r) + \frac{1}{2} < \rho_i \rho_k > \nabla_i \nabla_k V(r). \quad (6.39)$$

Obviously, the fluctuations $\boldsymbol{\rho}$ are isotropic, so that

$$< \rho_i >= 0, \quad < \rho_i \rho_k >= \frac{1}{3} < \rho^2 > \delta_{ik}.$$

Therefore, the perturbation of the potential by these fluctuations reduces to

$$\delta V(r) = \frac{1}{6} < \rho^2 > \Delta V(r). \quad (6.40)$$

For the attracting Coulomb potential $V(r) = - \alpha/r$ it gives

$$\delta V_C(r) = \frac{2\pi\alpha}{3} \delta(\mathbf{r}) < \rho^2 > \quad (6.41)$$

(here and below in this chapter we use mainly the units where $\hbar = 1$, $c = 1$, $\alpha = e^2$).

Let us consider now $< \rho^2 >$. The equation of motion of an electron in the fluctuating electric field $m\ddot{\mathbf{r}} = e\mathbf{E}$ gives for the Fourier components the following result:

$$\rho_\omega = - \frac{e\mathbf{E}_\omega}{m\omega^2}. \quad (6.42)$$

Due to the usual normalization condition, according to which the energy of vacuum fluctuations for given frequency and polarization is

$$\frac{\mathbf{E}_\omega^2 + \mathbf{B}_\omega^2}{8\pi} = \frac{\mathbf{E}_\omega^2}{4\pi} = \frac{\omega}{2},$$

we obtain

$$\rho_\omega^2 = \frac{2\pi\alpha}{m^2\omega^3}. \quad (6.43)$$

To obtain $< \rho^2 >$, we have to sum this expression by polarizations of the fluctuating field, i.e., multiply it by 2, and to integrate it over the phase space:

$$< \rho^2 > = 2 \int \frac{d\mathbf{k}}{(2\pi)^3} \, \rho_\omega^2 = \frac{2\alpha}{\pi m^2} \int \frac{d\omega}{\omega} . \tag{6.44}$$

With the logarithmic accuracy, we can take as a lower limit for this integral the binding energy $\sim m\alpha^2$: at smaller frequencies, the electron cannot be considered free, so that its motion due to the fluctuations is suppressed here. For the upper limit we accept $m$, since here the employed nonrelativistic approximation is not valid, the electron mass effectively grows, which also leads to the suppression of the motion due to the fluctuations. Thus, we arrive at the following expression for the Lamb perturbation in hydrogen:

$$\delta V_C(r) = \frac{8}{3} \frac{\alpha^2}{m^2} \ln \frac{1}{\alpha} \delta(\mathbf{r}) . \tag{6.45}$$

With this perturbation, we find with the logarithmic accuracy the known result for the Lamb shift in hydrogen:

$$\delta E_{nl} = \frac{8}{3} \frac{\alpha^2}{m^2} \ln \frac{1}{\alpha} |\psi_{nl}(0)|^2 = \frac{8m\alpha^5}{3\pi n^3} \ln \frac{1}{\alpha} \delta_{l0} ; \tag{6.46}$$

here $n$ and $l$ are the principal and orbital quantum numbers, respectively.

In particular, $2s_{1/2}$ level in hydrogen shifts up by

$$\delta E(2s_{1/2}) = \frac{m\alpha^5}{3\pi} \ln \frac{1}{\alpha} .$$

Thus, the well-known degeneracy of $2s_{1/2}$ and $2p_{1/2}$ levels in hydrogen is lifted. More accurate calculations in quantum electrodynamics (*N.M. Kroll, W.E. Lamb*, 1949; *J.B. French, V.F. Weisskopf*, 1949) give for the shift of $2s_{1/2}$ level the value

$$\delta E(2s_{1/2}) = \frac{m\alpha^5}{3\pi} \left( \ln \frac{1}{\alpha} - 1.089 \right) = 1034 \text{ MHz} ,$$

and for the splitting of $2s_{1/2}$ and $2p_{1/2}$ levels the value

$$E_{2s_{1/2}} - E_{2p_{1/2}} = 1057.91 \pm 0.01 \text{ MHz} ,$$

in complete agreement with the experimental result

$$1057.90 \pm 0.06 \text{ MHz} .$$

We note that in hydrogen-like ions the Lamb shift grows as $Z^4$. One power of $Z$ originates from the nonscreened Coulomb potential of the nucleus, and $Z^3$ from $|\psi(0)|^2$.

## 6.4.2 Lamb Shift and Infrared Divergence

In fact, the Lamb shift in hydrogen is directly related to the infrared divergence in the problem of the electron scattering on the Coulomb center [3]. We note that this scattering amplitude is described by the diagrams in Fig. 6.3. If the infrared divergence is regularized with the photon mass $\lambda$, the

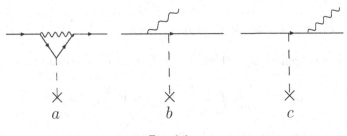

Fig. 6.3

logarithmic dependence of the vertex part (Fig. 6.3, $a$) on $\lambda$ cancels with the analogous dependence on $\lambda$ of diagrams in Figs. 6.3, $b$, $c$ which describe the Bremsstrahlung. (In the Coulomb gauge, used here and below, the dashed line refers to the Coulomb field, and the wavy line describes the transverse photon.)

If there is no acceleration, i.e., at the vanishing momentum transfer $q$, the radiation vanishes also. It is quite natural therefore that the infrared part of the vertex correction is also proportional to $q^2$. Indeed, with this correction included, the potential of the electron interaction with the Coulomb center is

$$V(\mathbf{q}) = - \frac{4\pi\alpha}{q^2} \left( 1 - \frac{\alpha q^2}{3\pi m^2} \ln \frac{m}{\lambda} \right). \qquad (6.47)$$

Of course, in the bound state problem there is no infrared radiation. However, here the electron is not free, it is off mass shell. Its deviation off the mass shell coincides in the order of magnitude with its binding energy, i.e., is estimated as $m\alpha^2$. On the other hand, the role of the photon mass in the Bremsstrahlung process is in essence to fix the minimum possible deviation of the total invariant mass of the final two-body state electron–photon from the mass of a free electron. Therefore, in the bound state problem one can put with the logarithmic accuracy in formula (6.47) $\lambda \sim m\alpha^2$. As a result, the logarithmic term in this formula reduces to

$$\delta V(\mathbf{q}) = \frac{8}{3} \frac{\alpha^2}{m^2} \ln \frac{1}{\alpha}. \qquad (6.48)$$

---

[3] In this and two next subsections we follow the work by *I. B. Khriplovich, A. I. Milshtein, A. S. Yelkhovsky* (1994).

Going over in this expression to the coordinate representation, we reproduce in such a way the Lamb interaction (6.45).

### 6.4.3 Lamb Shift in Two-Body Problem

Let us address now the case of two particles. The interaction potential for these particles, averaged over the vacuum fluctuations, is

$$< V(\mathbf{r}_1 - \mathbf{r}_2 + \boldsymbol{\rho}_1 - \boldsymbol{\rho}_2) >= V(\mathbf{r}_1 - \mathbf{r}_2) + \frac{1}{6} < (\boldsymbol{\rho}_1 - \boldsymbol{\rho}_2)^2 > \Delta V(\mathbf{r}_1 - \mathbf{r}_2) . \quad (6.49)$$

The arguments coinciding with the previous ones demonstrate that

$$< \rho_i^2 >= \frac{2e_i^2}{\pi m_i^2} \int_{m\alpha^2}^{m} \frac{d\omega}{\omega} = \frac{4e_i^2}{\pi m_i^2} \ln \frac{1}{\alpha} , \quad i = 1, 2 . \quad (6.50)$$

The corresponding contributions to the interaction operator are described by diagrams in Fig. 6.4; therein (and in diagrams below) the dashed line corresponds to the Coulomb interaction, and the wavy line describes the emission and absorption of the vacuum quantum.

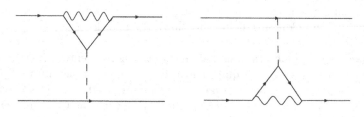

*Fig. 6.4*

Let us consider now the vacuum expectation value $2 < \boldsymbol{\rho}_1 \boldsymbol{\rho}_2 >$. It is distinct from zero only for fluctuations with the wavelength exceeding the size of the atomic system $1/(m\alpha)$ (here we also work with the logarithmic accuracy). For smaller wavelengths, or for larger frequencies, $\omega > m\alpha$, the fluctuations of coordinates are uncorrelated, i.e., $< \boldsymbol{\rho}_1 \boldsymbol{\rho}_2 >= 0$. In other words, the upper limit in the integral over frequencies of virtual quanta in the correlation function $< \boldsymbol{\rho}_1 \boldsymbol{\rho}_2 >$ is not $m$, as was the case in formula (6.44), but $m\alpha$. Thus, the contribution of this average is

$$-2 < \boldsymbol{\rho}_1 \boldsymbol{\rho}_2 >= - \frac{e_1 e_2}{\pi m_1 m_2} \int_{m\alpha^2}^{m\alpha} \frac{d\omega}{\omega} = - \frac{e_1 e_2}{\pi m_1 m_2} \ln \frac{1}{\alpha} . \quad (6.51)$$

It can be easily seen that the discussed vacuum expectation value corresponds to diagrams in Fig. 6.5.

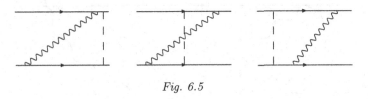

*Fig. 6.5*

So, the perturbation operator generated by diagrams in Figs. 6.4 and 6.5 is

$$\delta V_C^{e^+ e^-}(r) = 8 \frac{\alpha^2}{m^2} \ln \frac{1}{\alpha} \delta(\mathbf{r}) \qquad (6.52)$$

and

$$\delta V_C^{e^- e^-}(r) = -\frac{8}{3} \frac{\alpha^2}{m^2} \ln \frac{1}{\alpha} \delta(\mathbf{r}_{12}), \qquad (6.53)$$

correspondingly, for positronium and for atomic electrons (we recall that in the last case we deal with the Coulomb repulsion, but not attraction). At last, the operator for the corresponding interaction between electron and nucleus of mass $M$ and charge $Z$ is

$$\delta V_C(r) = \frac{8}{3} \frac{\alpha^2}{m^2} \ln \frac{1}{\alpha} \left(1 + \frac{Zm}{M}\right) \delta(\mathbf{r}_{12}); \qquad (6.54)$$

we confine ourselves here to the term of first order in small parameter $Zm/M$.

### 6.4.4 Magnetic Contribution and Thomson Scattering. Final Results for Two-Body Problems

Certainly, in the given order in $\alpha$ we have considered all contributions with the true infrared divergence which is cut off at the binding energy of the Coulomb system, i.e., at $m\alpha^2$. The above arguments demonstrate, however, that in diagrams with a double photon exchange (see Fig. 6.5) there is a contribution cutting them off effectively at frequencies larger than the typical momentum transfer $q \sim m\alpha$. It is natural therefore to consider in the same region $m\alpha < \omega < m$ the diagrams with double magnetic exchange. To our accuracy one can neglect in them the three-dimensional external momenta of both particles. It is well-known that in this case, in the totally nonrelativistic limit, the scattering of a transverse photon is described by the contact operator (6.22). Correspondingly, the double magnetic exchange is reduced to the simple diagrams (see Fig. 6.6) with the vertices

$$\frac{e_i^2}{m_i} \sqrt{\frac{4\pi}{2\omega}} \sqrt{\frac{4\pi}{2\omega'}} \left(\mathbf{e}^\lambda(\mathbf{k}) \, \mathbf{e}'^{\lambda'}(\mathbf{k}')\right), \qquad (6.55)$$

generated by this operator. Here $\omega$, $\omega'$ are the frequencies of the emitted quanta, $\mathbf{k}$, $\mathbf{k}'$ are their momenta, and $\mathbf{e}^\lambda(\mathbf{k})$, $\mathbf{e}'^{\lambda'}(\mathbf{k}')$ are their polarization

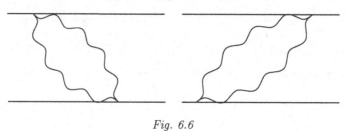

Fig. 6.6

vectors. Since the recoil of the particle – source is neglected, then in the present case

$$\mathbf{k}' = -\mathbf{k}, \quad \omega' = \omega, \quad \mathbf{e}^\lambda(\mathbf{k})\,\mathbf{e}'^{\lambda'}(\mathbf{k}') = \mathbf{e}^\lambda(\mathbf{k})\,\mathbf{e}^{\lambda'}(-\mathbf{k}) = \delta_{\lambda\lambda'}.$$

As a result, vertex (6.55) simplifies to

$$\frac{e_i^2}{m_i}\frac{4\pi}{2\omega}\delta_{\lambda\lambda'}. \tag{6.56}$$

The calculation of effective interaction $\delta V_M$ is of no difficulty. The common perturbation theory gives the following result in the momentum representation:

$$\delta V_M = 2 \cdot \frac{1}{2} \cdot \frac{e_1^2 e_2^2}{m_1 m_2} \sum_{\lambda\lambda'} \int \frac{d\mathbf{k}}{(2\pi)^3} \left(\frac{4\pi}{2\omega}\right)^2 \frac{\delta_{\lambda\lambda'}\,\delta_{\lambda\lambda'}}{-2\omega}. \tag{6.57}$$

Here, the overall factor 2 reflects the presence of two diagrams (see Fig. 6.6); the factor $1/2$ is due to the identity of photons: to avoid double count, the standard two-particle phase space for identical particles should be divided by two. And at last, $-2\omega$ in this formula is the common energy denominator arising in the second order of perturbation theory.

We note now that with two independent polarizations of the three-dimensionally transverse photons

$$\sum_{\lambda\lambda'}\delta_{\lambda\lambda'}\,\delta_{\lambda\lambda'} = \sum_\lambda \delta_{\lambda\lambda} = 2.$$

Then we recall that $k = \omega$, and that the effective region of integration over $\omega$ extends from $m\alpha$ to $m$. Thus, we obtain

$$\delta V_M = -2\frac{e_1^2 e_2^2}{m_1 m_2} \int_{m\alpha}^m \frac{d\omega}{\omega} = -2\frac{e_1^2 e_2^2}{m_1 m_2} \ln\frac{1}{\alpha},$$

or in the coordinate representation

$$\delta V_M = -2\frac{e_1^2 e_2^2}{m_1 m_2} \ln\frac{1}{\alpha}\,\delta(\mathbf{r}_{12}). \tag{6.58}$$

For the electron–electron and electron–positron interactions, this expression reduces to

$$-2\frac{\alpha^2}{m^2}\ln\frac{1}{\alpha}\delta(\mathbf{r}_{12}),\tag{6.59}$$

and for the electron–nucleus interaction to

$$-2\frac{Z^2\alpha^2}{mM}\ln\frac{1}{\alpha}\delta(\mathbf{r}).\tag{6.60}$$

The total Lamb shift operator in positronium is

$$\delta V(\mathbf{r}) = \delta V_C(\mathbf{r}) + \delta V_M(\mathbf{r}) = 6\frac{\alpha^2}{m^2}\ln\frac{1}{\alpha}\delta(\mathbf{r}),\tag{6.61}$$

and the corresponding shift of a level with quantum numbers $n$, $l$ constitutes

$$\delta E_{nl} = 6\frac{\alpha^2}{m^2}\ln\frac{1}{\alpha}|\psi_{nl}(0)|^2 = \frac{3}{4}\frac{m\alpha^5}{\pi n^3}\ln\frac{1}{\alpha}\delta_{l0}.\tag{6.62}$$

This result reproduces the logarithmic contribution to the exact expression for the Lamb shift in positronium (*T. Fulton, P.C. Martin,* 1954). Numerical difference between the logarithmic result (6.62) and the exact one is small for parapositronium (the total spin $S = 0$): instead of $\ln\alpha = 4.9$, we have in the exact result 4.7. The difference in orthopositronium ($S = 1$) is larger: $\ln\alpha = 4.9$ changes in the exact result to 3.0.

In the presented calculation of the logarithmic contribution to the Lamb shift, it becomes quite clear why this contribution is independent of $S$. For the correction $\delta V_C(\mathbf{r})$, this is the spin independence of the common Coulomb interaction. For the correction $\delta V_M(\mathbf{r})$, related to the double magnetic exchange, this is the spin independence of the Thomson amplitude, i.e., of the nonrelativistic limit of the Compton scattering.

Let us come back to usual atoms. With formulas (6.53), (6.54), (6.59), and (6.60), we obtain the total Lamb shift operator for an atom, including the contribution of first order in $1/M$:

$$\delta V = \frac{8}{3}\frac{Z\alpha^2}{m^2}\ln\frac{1}{\alpha}\sum_i\delta(\mathbf{r}_i)$$

$$-\frac{14}{3}\frac{\alpha^2}{m^2}\ln\frac{1}{\alpha}\sum_{i<j}\delta(\mathbf{r}_{ij}) + \frac{2}{3}\frac{Z^2\alpha^2}{mM}\ln\frac{1}{\alpha}\sum_i\delta(\mathbf{r}_i).\tag{6.63}$$

The electron–electron interaction in this expression was known for helium (*H. Araki,* 1957; *P.K. Kabir, E.E. Salpeter,* 1957; *J. Sucher,* 1958), and the term $\sim 1/M$ was known for hydrogen (*G.W. Erickson, D.R. Yennie,* 1965).

Thus, all logarithmic terms in the Lamb shift have clear physical interpretations.

## 6.5 High-Energy Processes

### 6.5.1 Does Classical Limit Exist for Electron Scattering in Coulomb Field?

The answer to this question for the nonrelativistic problem is well-known: certainly, here this limit does exist. Moreover, the classical cross-section for the electron scattering on the Coulomb center in the nonrelativistic case just coincides with the quantum one [4].

The situation in the relativistic problem is more subtle. The scattering cross-section for an electron on the Coulomb center with charge $Ze$ is

$$d\sigma_{1/2} = d\sigma_0 \left(1 - v^2 \sin^2 \frac{\theta}{2}\right), \tag{6.64}$$

where

$$d\sigma_0 = \frac{(Z\alpha)^2}{4\varepsilon^2 v^4} \sin^{-4} \frac{\theta}{2} \, d\Omega \tag{6.65}$$

is the scattering cross-section for particle of spin 0 on the same center, $\varepsilon$ and $v$ are the energy and velocity of the particle, and $\theta$ is the scattering angle.

Due to the factor

$$\left(1 - v^2 \sin^2 \frac{\theta}{2}\right) \tag{6.66}$$

in formula (6.64), the backward scattering of the electron, at $\theta \to \pi$, in the ultrarelativistic limit $v \to 1$ is suppressed. The explanation of this suppression is sufficiently simple. In the ultrarelativistic limit the electron helicity (i.e., the projection of the spin onto the momentum) is conserved in electromagnetic interactions. Since the projection of the orbital angular momentum onto the momentum vanishes identically, the helicity conservation means that the projection of the total angular momentum onto the momentum is conserved. Thus, the backward scattering of ultrarelativistic electron is forbidden by the conservation of the total angular momentum; in the present case we mean the conservation of its projection onto the direction of the initial motion.

However, the natural question arises. Of course, the spin of a particle is its quantum characteristic. Then, how is it that scattering cross-sections (6.64) and (6.65) for particles of spin 1/2 and 0, respectively, differ by suppression factor (6.66) which does not contain the Planck constant $\hbar$? The usual answer is that both formulas, (6.64) and (6.65), have been derived in fact in the Born approximation, under the condition $Ze^2/\hbar v \ll 1$, and therefore do not allow the limiting transition $\hbar \to 0$.

Of course, this is true by itself, but in no way exhausts the problem. Indeed, the helicity of ultrarelativistic electron is conserved beyond the Born

---

[4]Let us recall that once this coincidence proved to be extremely important for the correct interpretation of the Rutherford experiments and for the creation of quantum mechanics.

approximation as well, so that its backward scattering is suppressed as compared to the scalar case in the exact problem also. On the other hand, in the classical limit $\hbar \to 0$, one may expect that the difference between particles of spin $1/2$ and $0$ is erased. So, how does the transition to the classical limit occur in the exact solution?

The answer to this question is rather unexpected. The problem of scattering on a Coulomb center does not allow the limiting transition $\hbar \to 0$ for both spins, $1/2$ and $0$. It is well-known that the ground state energy of electron in the field of a point-like nucleus with charge $Z$ is [5]

$$E = mc^2 \sqrt{1 - \left(\frac{Ze^2}{\hbar c}\right)^2} \; ; \tag{6.67}$$

we write down explicitly in this expression not only $\hbar$, but $c$ as well. It is quite obvious that formula (6.67) does not allow the limiting transition $\hbar \to 0$. But what happens really at $Ze^2/\hbar c \to 1$? At first, the total energy of electron tends to zero. In the realistic problem, one should take into account the finite size of the nucleus. Then, with the further decrease of $\hbar$, or (which is the same) with the increase of $Z$, this energy becomes negative. With the subsequent increase of $Z$, it becomes negative to such an extent that the bare nucleus can decay into a single-electron ion and positron (*V.N. Gribov*, 1974). Of course, this reasoning applies as well to a Coulomb center with a negative charge which will decay into electron and the bound state of positron.

Let us come back to our problem. These arguments demonstrate that with such decrease of $\hbar$, under which $Ze^2/\hbar c$ tends to 1, the very approximation of an external Coulomb field loses physical meaning.

### 6.5.2 Cross-Sections of Processes $e^+e^- \to \mu^+\mu^-$ and $e^+e^- \to \pi^+\pi^-$

The processes $e^+e^- \to \mu^+\mu^-$ and $e^+e^- \to \pi^+\pi^-$ are described by Feynman diagrams presented in Fig. 6.7. Their total cross-sections are well-known, and in the center-of-mass frame are, respectively,

$$\sigma_\mu = \frac{4\pi\alpha^2}{3\varepsilon^2}\left(1 - \frac{m_\mu^2}{\varepsilon^2}\right)^{1/2}\left(1 + \frac{m_\mu^2}{2\varepsilon^2}\right) \tag{6.68}$$

and

$$\sigma_\pi = \frac{\pi\alpha^2}{3\varepsilon^2}\left(1 - \frac{m_\pi^2}{\varepsilon^2}\right)^{3/2} . \tag{6.69}$$

---

[5]Let us note that for a charged scalar particle the corresponding ground state energy is

$$E = mc^2 \sqrt{1/2 + \sqrt{1/4 - (Ze^2/\hbar c)^2}} .$$

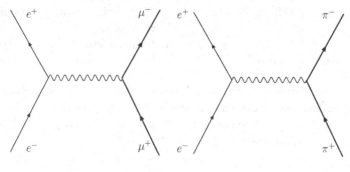

*Fig. 6.7*

Here $\varepsilon$ is the energy of initial electron, $m_\mu$ and $m_\pi$ are the muon and pion masses. It does not make sense to present here the standard calculation of these processes which can be found in numerous textbooks. However, some curious, and sometimes unexpected, peculiarities of these reactions are of a doubtless interest. We discuss them below, trying to confine ourselves to qualitative arguments.

Let us start with the current conservation law which in the momentum representation is $q_\mu j_\mu = 0$, where the momentum transfer $q_\mu = p'_\mu - p_\mu$ in the center of mass system is $(2\varepsilon, 0)$. (Let us recall that here the 4-momentum of an antiparticle with energy $\varepsilon$ and momentum $\mathbf{p}$ should be understood as $p_\mu = -(\varepsilon, \mathbf{p})$.) It means in particular that the currents of both initial and final particles, as well as the intermediate photon, have the space components only. Hence, both pairs of particles, initial and final, are in a state with the total angular momentum $J = 1$.

For the reaction $e^+ e^- \to \pi^+ \pi^-$, the quantum numbers of the final state follow from it immediately: $^1P_1$ (we use the standard notations $^{2S+1}L_J$, where $S$ is the total spin of the system, $L$ is its total orbital angular momentum, $J$ is its total angular momentum). It is well-known that the amplitude of a two-particle process at the threshold, i.e., at $p \to 0$, is proportional to $p^L$, and its cross-section is proportional to $p^{2L+1}$ (one power of the momentum originates from the two-particle phase space). Thus, the threshold behavior of cross-section (6.69),

$$ \sigma_\pi \sim \left( 1 - \frac{m_\pi^2}{\varepsilon^2} \right)^{3/2} \sim p^3 , $$

is quite obvious. We note that the space parity of a system of two (pseudo)scalar particles $\pi^+\pi^-$, equal to $(-1)^L$, or $-1$ in our case, coincides, as it should, with the space parity of the intermediate photon.

We go over now to the muon production. Let us determine the quantum numbers $L$ and $S$, possible here at the total angular momentum $J = 1$. The space parity of the fermion – antifermion state is equal to $(-1)^{L+1}$, so that only even $L$ are possible here. With the condition $J = 1$ (and the obvious limitation $S \leq 1$), we find two possible states of the system $\mu^+\mu^-$: $^3S_1$ and $^3D_1$. Let us note that in the result the charge parity of this system which is well-known to be equal to $(-1)^{L+S}$, has coincided, as should be expected, with the charge parity $-1$ of the intermediate photon. Of course, the total cross-section at the threshold is dominated by the contribution of the $^3S_1$ state, so that in this case as well the threshold behavior,

$$\sigma_\mu \sim \left(1 - \frac{m_\mu^2}{\varepsilon^2}\right)^{1/2} \sim p,$$

is quite natural.

Let us discuss now the asymptotic behavior of the cross-sections $\sigma_\mu$ and $\sigma_\pi$ for $\varepsilon \gg m_{\mu,\pi}$. By dimensional reasons, it is quite natural that both cross-sections decrease as $1/\varepsilon^2$ at $\varepsilon \to \infty$ (we recall that the coupling constant $\alpha$ is dimensionless). Perhaps, more curious is the problem of the asymptotic ratio of cross-sections: $\sigma_\mu/\sigma_\pi \to 4$ at $\varepsilon \to \infty$.[6] Below we will be engaged in the analysis of this asymptotic.

We start with a more detailed discussion of the structure of the electromagnetic currents, boson and fermion ones. For $\pi$-mesons it is simple and obvious:

$$J_\mu^\pi = (0, 2\,\mathbf{p}), \tag{6.70}$$

i.e., the total charge density of the particle and antiparticle in the center of mass system vanishes, and the total current density is directed along the momentum of one of them.

We go over now to fermions. Here the total charge density $J_0$ of course vanishes as well, and the three-dimensional current density is $\mathbf{J} = u^\dagger \boldsymbol{\alpha} v$, where

$$\alpha = \begin{pmatrix} 0 & \sigma \\ \sigma & 0 \end{pmatrix};$$

and the bispinors

$$u^\dagger = \left(\phi^\dagger \sqrt{\varepsilon + m_\mu},\ \phi^\dagger \sqrt{\varepsilon - m_\mu}\,(\boldsymbol{\sigma}\mathbf{n})\right) \tag{6.71}$$

and

$$v = \begin{pmatrix} -\sqrt{\varepsilon - m_\mu}\,(\boldsymbol{\sigma}\mathbf{n})\,\chi \\ \sqrt{\varepsilon + m_\mu}\,\chi \end{pmatrix} \tag{6.72}$$

describe the created $\mu^-$ and $\mu^+$ (or annihilating $e^+$ and $e^-$). In expressions (6.71) and (6.72), $\phi$ and $\chi$ are the corresponding two-component spinors normalized to unity, and $\mathbf{n} = \mathbf{p}/p$. We note that, with bispinors (6.71) and (6.72),

---

[6]The discussion of this problem looks proper since from time to time, at least in folklore, one comes across absolutely absurd attempts to explain this ratio.

$J_0 = 0$, as it should be. As to the expression for the three-dimensional current (we recall that here and below the annihilation current is discussed), it reduces with (6.71) and (6.72) to

$$\mathbf{J} = 2\,\phi^\dagger\left[\varepsilon\,\boldsymbol{\sigma} - (\varepsilon - m_\mu)\,\mathbf{n}(\boldsymbol{\sigma}\mathbf{n})\right]\chi$$

$$= \frac{2}{3}\,\phi^\dagger\left\{(2\,\varepsilon + m_\mu)\,\boldsymbol{\sigma} - (\varepsilon - m_\mu)\left[3\,\mathbf{n}(\boldsymbol{\sigma}\mathbf{n}) - \boldsymbol{\sigma}\right]\right\}\chi. \qquad (6.73)$$

In the last line of this expression we have singled out the vector structure $3\,\mathbf{n}(\boldsymbol{\sigma}\mathbf{n}) - \boldsymbol{\sigma}$, which contains the irreducible second-rank space tensor $3\,n_i n_j - \delta_{ij}$ and hence describes the $^3D_1$ amplitude. The vector left, just $\boldsymbol{\sigma}$, is independent of $\mathbf{n}$ and therefore refers to the $^3S_1$ amplitude. Of course, in the total cross-section, the amplitudes $^3S_1$ and $^3D_1$ do not interfere.

We note now that the vector of fermion current (6.73), as distinct from the $\pi$-meson one (6.70) (and from naïve expectations!), in no way is directed along $\mathbf{n}$, the line of flight of the particles. Moreover, in the ultrarelativistic limit this vector of current,

$$\mathbf{J} = 2\,\varepsilon\phi^\dagger\left[\boldsymbol{\sigma} - \mathbf{n}(\boldsymbol{\sigma}\mathbf{n})\right]\chi \qquad (6.74)$$

(see the first line of equation (6.73)), lies completely in the plane orthogonal to $\mathbf{n}$.

The explanation of this, rather unexpected circumstance is as follows. For particles with magnetic moment, the electromagnetic current contains, in line with the common, convection contribution, an additional, spin-dependent term. For instance, in the usual nonrelativistic problem of electron scattering on a Coulomb center, this additional current is directed along $\boldsymbol{\sigma} \times \mathbf{q}$, where $\boldsymbol{\sigma}$ is the vector of spin, and $\mathbf{q}$ is the momentum transfer. In our case, which is the annihilation of ultrarelativistic spinning particles, this additional term cancels completely the usual convection contribution, so that the resulting electron current flows only in the plane orthogonal to the line of collision of the beams, and the current of final muons flows in the plane orthogonal to their line of flight.

The cancellation between the convection and spin contributions to the annihilation fermion current can also be demonstrated as follows. Let us rewrite the standard expression for this current in this way:

$$\bar{u}(p')\gamma_\mu u(p) = \frac{1}{2m}\,\bar{u}(p')\left[(p'_\mu + p_\mu) + i\sigma_{\mu\nu}q_\nu\right]u(p), \qquad (6.75)$$

where

$$\sigma_{\mu\nu} = \frac{i}{2}(\gamma_\mu\gamma_\nu - \gamma_\nu\gamma_\mu), \quad q_\nu = p'_\nu - p_\nu.$$

Obviously, the structure $\bar{u}(p')(p'_\mu + p_\mu)u(p)/2m$ in this expression corresponds to the convection current, and $\bar{u}(p')i\sigma_{\mu\nu}q_\nu u(p)/2m$ corresponds to the spin one. The necessity of strong cancellation between them becomes obvious at least because each of these terms grows with energy more rapidly than the initial Dirac expression $\bar{u}(p')\gamma_\mu u(p)$.

To investigate the problem further, it is convenient to drop in each three-dimensional current of ultrarelativistic particles the overall factor $2\varepsilon$, thus making these currents dimensionless. In this way, choosing as the $z$ axis for each of the currents the unit vectors of the momenta of created $\pi^-$, $\mu^-$, and annihilating $e^-$, respectively, we find

$$\mathbf{J}^\pi = 2\mathbf{p} \to \mathbf{n} = (0,0,1), \tag{6.76}$$

$$\mathbf{J}^\mu = 2\varepsilon\phi^\dagger\left[\boldsymbol{\sigma} - \mathbf{n}(\boldsymbol{\sigma}\mathbf{n})\right]\chi \to \phi^\dagger\left[\boldsymbol{\sigma} - \mathbf{n}(\boldsymbol{\sigma}\mathbf{n})\right]\chi = (1,\pm i,0) \tag{6.77}$$

($+i$ and $-i$ refer here to the created $\mu^-$ of left and right helicity, respectively),

$$\mathbf{J}^e = 2\varepsilon\phi^\dagger\left[\boldsymbol{\sigma} - \mathbf{n}(\boldsymbol{\sigma}\mathbf{n})\right]\chi \to \phi^\dagger\left[\boldsymbol{\sigma} - \mathbf{n}(\boldsymbol{\sigma}\mathbf{n})\right]\chi = (1,\mp i,0) \tag{6.78}$$

($-i$ and $+i$ refer here to the annihilating electrons of left and right helicity, respectively).

We start with the angular distribution of the final particles in the reaction $e^+e^- \to \pi^+\pi^-$. Its amplitude is proportional to $(\mathbf{J}^\pi\mathbf{J}^e) = (\mathbf{n}_\pi\mathbf{J}^e)$. If the angle between the momenta of $\pi$-mesons and the momenta of ultrarelativistic colliding beams is $\theta$, then the angle between the momenta of $\pi$-meson and the current $\mathbf{J}^e$, which lies in the plane orthogonal to the momenta of the colliding beams, is $\pi/2 - \theta$. Choosing the projection of $\mathbf{n}^\pi$ onto this plane as the $x$ axis on it, we find immediately that $(\mathbf{n}^\pi\mathbf{J}^e) = \cos(\pi/2 - \theta) = \sin\theta$. Correspondingly, the angular distribution of the pions is

$$d\sigma_\pi \sim \sin^2\theta = 1 - \cos^2\theta. \tag{6.79}$$

Now we address the angular distribution of muons in the process $e^+e^- \to \mu^+\mu^-$. Here it is convenient to choose the line of intersection of the planes orthogonal to $\mathbf{n}^e$ and $\mathbf{n}^\mu$, as the $y$ axis, common to both planes. Then one can easily see that

$$(\mathbf{J}^\mu\mathbf{J}^e) = \cos\theta \pm 1. \tag{6.80}$$

Here $\theta$ is the angle between $\mathbf{n}^e$ and $\mathbf{n}^\mu$, and the signs $+$ or $-$ before 1 arise, respectively, for same or opposite helicities of $e^-$ and $\mu^-$. We note that the negative sign for opposite helicities in expression (6.80) is quite natural: the conservation of projection of the total angular momentum forbids the forward scattering (more precisely, creation) with the helicity flip. And at last, the angular distribution of muons, after the summation over their helicities, is

$$d\sigma_\mu \sim (1 + \cos\theta)^2 + (1 - \cos\theta)^2 = 2(1 + \cos^2\theta). \tag{6.81}$$

Now the origin of the asymptotic ratio of the total cross-sections, $\sigma_\mu/\sigma_\pi = 4$, becomes clear. Of course, one factor 2 is caused by the larger statistical weight for the muons, i.e., by the presence of two helicities for each of them (see (6.81)); it should be kept in mind here that these helicities are correlated:

one of the muons is always left, and another is right. One more factor 2 is due to different angular distributions of pions and muons: the integral over the total solid angle of $1 + \cos^2 \theta$ is twice as large as that of $\sin^2 \theta$ (see (6.79) and (6.81)).

# Index